This book is to be returned on or before
the last date stamped below.

CONTROLLED AIR INCINERATION

\# 12635705

Controlled Air Incineration

FRANK L. CROSS, JR., PE
HOWARD E. HESKETH, PHD, PE

LANCASTER · BASEL

UNIVERSITY OF
STRATHCLYDE LIBRARIES

Published in the Western Hemisphere by
Technomic Publishing Company, Inc.
851 New Holland Avenue
Box 3535
Lancaster, Pennsylvania 17604 U.S.A.

Distributed in the Rest of the World by
Technomic Publishing AG

©1985 by Technomic Publishing Company, Inc.
All rights reserved

No part of this publication may be reproduced, stored in a retrieval system, or transmitted, in any form or by any means, electronic, mechanical, photocopying, recording, or otherwise, without the prior written permission of the publisher.

Printed in the United States of America
10 9 8 7 6 5 4 3 2 1

Main entry under title:
 Controlled Air Incineration

A Technomic Publishing Company book
Bibliography: p.
Includes index p. 113

Library of Congress Card No. 85-51123
ISBN No. 87762-396-1

TABLE OF CONTENTS

Acknowledgments ix

1 Incinerators and Combustion   1

    1.1   Introduction   1

    1.2   Controlled Air Incinerators   3

    1.3   Combustible Wastes   5
            1.3.1   Volatile Matter   7
            1.3.2   Fixed Carbon   8
            1.3.3   Moisture   8
            1.3.4   Noncombustibles   8
            1.3.5   Classification of Wastes   9

    1.4   Combustion   9
            1.4.1   Types of Combustion   9
            1.4.2   Theory of Combustion   9
            1.4.3   Excess Air   14
            1.4.4   Air-Fuel Ratio   14
            1.4.5   Equivalence Ratio   15
            1.4.6   Factors that Affect the Combustion Process   15
            1.4.7   Devolatilization Characteristics   16
            1.4.8   Combustion Simplified   16
            1.4.9   Flue Gas Composition   18
            1.4.10 Flue Gas Quantity   20

    1.5   Incinerator Types   34

2 Incinerator Operation   35

    2.1   Temperature Control and Excess Air   35

2.2   Loading   35

2.3   System Behavior   41

2.4   Air Control   41
    2.4.1   Manual Air Control   42
    2.4.2   Oxygen Modulated Air Control   42
    2.4.3   Temperature Modulated Air Control   42

2.5   Operation by Visual Observation   45

2.6   Auxiliary Fuel Usage   46

2.7   Clinkers   47

2.8   Refractory Considerations   48
    2.8.1   General   48
    2.8.2   Refractory Types   48
    2.8.3   Chemical, Thermal and Corrosion Concerns   49
    2.8.4   Refractory Selection   50
    2.8.5   Refractory Costs   51
    2.8.6   Installation Time   53

2.9   Burners   53
    2.9.1   Gas Burners   54
    2.9.2   Oil Burners   54
    2.9.3   Combination Gas-Oil Burners   55
    2.9.4   Ignition Systems   55
    2.9.5   Burner Controls and Safety Equipment   56
    2.9.6   Burner Rate Checks   57

2.10   Spare Parts   57

2.11   Design Considerations   58

2.12   Problem Waste Operation   59

2.13   Maintenance   61
    2.13.1   Typical Maintenance Procedures   62
    2.13.2   Troubleshooting   63

2.14   Routine Operation   65

3 Emissions, Controls, Energy Recovery and Costs   69

3.1   Incinerator Emissions   69
    3.1.1   Stack Emissions   69
    3.1.2   Emission Factors   71

3.1.3 Particle Size Distribution 71
 3.1.4 Air Pollution Control 71
 3.1.5 Emission Testing 73
 3.1.6 Burning and Testing of Hazardous Wastes 74
 3.1.7 Sampling and Analytical Procedures for Organic Emissions 75

3.2 Energy Recovery 88
 3.2.1 Heat Recovery 89
 3.2.2 Material Recovery 92
 3.2.3 Design Effects 95

3.3 Costs 95
 3.3.1 Incineration Systems 95
 3.3.2 Air Pollution Control Systems 96

Appendix 107

Fans and Dampers 107

Conversion Factors 111

Lower Explosive Limits in Air 111

Index 113

Biographies 115

ACKNOWLEDGMENTS

The authors wish to thank Robert Marcinko and Yun-Yu Liu for reviewing the manuscript and indexing this book.

CHAPTER 1

Incinerators
AND
Combustion

1.1 INTRODUCTION

Incinerators have been used for centuries to dispose of combustible wastes from industry, commercial applications and the public. Up until the mid 1950's, they were considered a necessary evil and initially this activity was undertaken in the cheapest possible manner. However as air pollution regulations became more stringent and as the waste heat recovery aspects became more attractive, incineration systems were improved dramatically.

There are four principle types of incinerators used for the disposal of solid waste—open burning, single chamber, multiple chamber, and controlled air incinerators. Los Angeles County passed regulations banning open burning and single chamber incinerators as early as September 1957 [1]. Passage of the Resource Conservation and Recovery Act (RCRA) in 1976 spurred on further incinerator development work. Since then the controlled air incinerator has become more and more widely used because of its ability to meet increasingly stringent air pollution control standards. A controlled air incinerator is simply a multiple chamber unit with the capabilities for very careful control of air flow rates and injection positions as discussed in Section 1.2. The controlled air incinerator is characterized by very low particulate emissions (see Table 1.1).

The evolution of the controlled air incinerator is closely related to the growth of air pollution control standards and regulations. Early incinerators were of the single chamber "excess air" type, in which a high air flow was used for not only combustion, but also for cooling of the grate. While frequently used in large municipal waste disposal facilities, such units can be of almost any size and capacity.

The use of excess air to cool the grate generally produces considerable

2 CONTROLLED AIR INCINERATION

TABLE 1.1
Average controlled air incinerator emission factors for refuse combustion.

	Emissions, lb/ton			
	Particulates	CO	HC	NOx
Open Burning*	16	85	30	6
Single Chamber**	15	20	15	2
Multiple Chamber**	7	10	3	3
Controlled Air**	1.4	Neg.	Neg.	10

*Municiple refuse.
**Industrial—commercial units.
Neg—Negligible.

turbulence within the combustion chamber, resulting in relatively high particulate emissions. Thus, as clean air standards became more stringent, particulate emission controls were required. At first, it was found that a significant improvement in particulate emission control could be made by passing the combustion products from the single chamber incinerator through a secondary combustion chamber as in Figure 1.1.

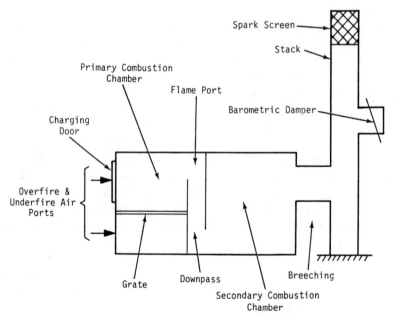

FIGURE 1.1 *Typical parts of an in-line multiple chamber incinerator.*

Another radical alteration spurred on by the Toxic Substances Control Act (TSCA) in 1976 was the requirement for 1200 °C (2200 °F) combustion temperature and greater than 3% excess oxygen to incinerate polychlorinated biphenyl compounds (PCB's).

The secondary chamber in the smaller in-line incinerators act as a simple gas-treating device by permitting the settlement of large particles in the gas stream from the primary combustion chamber. Table 1.1 indicates that these multiple chamber units did make a significant improvement over single chamber units in the reduction of air pollution from the combustion of refuse. Additional improvements were made on these units by adding auxiliary heat for both the primary and secondary combustion chambers to provide optimum temperatures for complete combustion. However, erratic air supply control frequently negated the benefits of these auxiliary heating units, since it was too expensive and impractical to heat the very large quantities of air which passed through the incinerators.

For most applications where general wastes were burned, excess air was drawn into the unit by a natural stack draft (often accompanied by a barometric damper). In special cases, such as the burning of plastics or other uniform dry wastes with a high BTU content, forced draft was frequently provided.

In 1981 requirements were proposed to require one or more air pollution control devices for every hazardous waste incinerator. As incinerator air pollution emission regulations became more stringent, scrubbing control devices were added to the units to remove additional particulates from the discharge gases as in Figure 1.2a. These devices are more effective in removing large quantities of particulate pollutants. However, many are not entirely successful in meeting the more stringent air pollution emission standards.

1.2 CONTROLLED AIR INCINERATORS

The next step in meeting more and more stringent air pollution control standards was the development of the "controlled air" incinerator. This type of incinerator consists of two chambers—an ignition or primary chamber and an upper secondary or combustion chamber as shown in Figure 1.2b. Generally, the lower chamber is operated at less-than-stoichiometric air conditions[1] to minimize particulate entrainment. This

---
[1]Less than the amount of air needed for complete combustion.

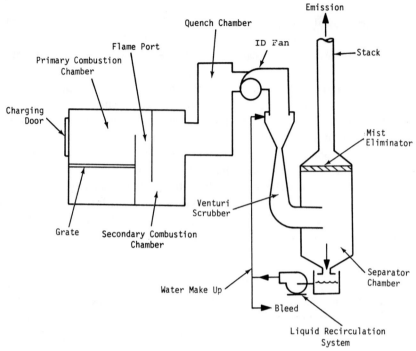

FIGURE 1.2a *In-line incinerator with wet scrubber.*

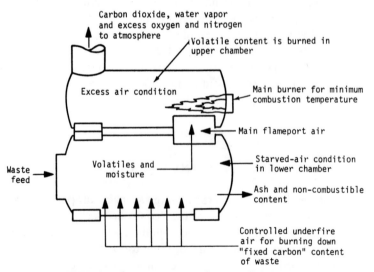

FIGURE 1.2b *Component parts of a controlled air incinerator.*

chamber serves primarily as a volatilization chamber. The gases produced in the lower chamber are drawn into the secondary chamber where excess air is injected under turbulent conditions to achieve final combustion. Gas retention time in the upper chamber is on the order of 0.25 to 2.0 seconds dependent upon the waste chemistry, manufacturer, and specification requirements. Controlled air incinerators are able to dispose of a wide variety of organic solid wastes with extremely low particulate emissions as presented in Table 1.1.

1.3 COMBUSTIBLE WASTES

Wastes come in many types, sizes and colors. Some typical wastes are listed in Table 1.2 with the moisture ash free (MAF) heating value and some typical compositions. Note that certain wastes can have very high heating values. For example, rubber is about 12,500 Btu/lb and suet is about 16,200 Btu/lb [2]. As indicated in Table 1.2, the chemical composition of waste is described by a "proximate analysis," which is simply a laboratory determination of four major components of the waste material. These components consist of the following:

Wt. % Volatile Matter
Wt. % Fixed Carbon
Wt. % Moisture
Wt. % Ash or Noncombustibles
100 % Total

Volatile matter is the component of the waste which can be liberated by the application of heat only. In theory, the volatile matter is burned in the secondary chamber, upon exposure to secondary combustion air. In practice, some of the volatile matter is usually burned in the primary chamber, but this is held to a minimum. Volatile matter combustion is a gas-phase reaction.

Fixed carbon is the nonvolatile component of the waste, which is burned at higher temperatures and at longer exposure to combustion air. The combustion of fixed carbon is a solid-phase reaction.

Moisture is evaporated from the waste due to heat in the primary chamber. It passes through the secondary chamber and out of the incinerator as superheated waste vapor.

Ash and noncombustibles are the components of the waste which are not burned, and remain in the primary chamber. This component causes

TABLE 1.2
Higher heating value and proximate analyses of various wastes.

		Typical Weight Percent			
Components	Approx. Btu/lb	Moisture	Volatile Matter	Fixed Carbon	Non-Combustibles
Paper and Paper Products	7,900				
Paper mixed		10.24	75.94	8.44	5.38
Newsprint		5.97	81.12	11.48	1.48
Brown Paper		5.83	83.92	9.24	1.01
Trade Magazine		4.11	66.39	7.03	22.47
Corrugated Bones		5.20	77.47	12.27	5.06
Plastic-Coated Paper		4.71	84.20	8.45	2.64
Waxed Milk Cartons		3.45	90.92	4.46	1.17
Paper Food Cartons		6.11	75.59	11.80	6.50
Junk Mail		4.56	73.32	9.03	13.09
Food and Food Waste	14,000–16,000				
Vegetable Food Waste		78.29	17.10	3.55	1.06
Citrus Rinds and Seeds		78.70	16.55	4.01	0.74
Meat Scraps Cooked		38.74	56.34	1.81	3.11
Fried Fats		0.00	97.64	2.36	0.00
Trees, Wood, Plants	7,500–8,500				
Green Logs		50.00	42.25	7.25	0.50
Rotten Timbers		26.80	55.01	16.13	2.06
Demolition Softwood		7.70	77.62	13.93	0.75
Waste Hardwood		12.00	75.05	12.41	0.54
Furniture Wood		6.00	80.92	11.74	1.34
Evergreen Shrubs		69.00	25.18	5.01	0.81
Balsam Spruce		74.35	20.70	4.13	0.82
Flowering Plants		53.94	35.64	8.08	2.34
Lawn Grass		75.24	18.64	4.50	1.62
Ripe Leaves I		9.97	66.92	19.29	3.82
Ripe Leaves II		50.00	—	—	4.10
Wood and Bark		20.00	67.89	11.31	0.80
Mixed Greens		62.00	26.74	6.32	4.94
Domestic Wastes	7,000–7,500				
Upholstery		6.9	75.96	14.52	2.62
Tires		1.02	64.92	27.51	6.55
Leather		10.00	68.46	12.49	9.10
Leather Shoe		7.46	57.12	14.26	21.16
Shoe Heel and Sole		1.15	67.03	2.08	29.74
Rubber		1.20	83.98	4.94	9.88
Mixed Plastics		2.00	—	—	10.00
Polyethylene		0.20	98.54	0.07	1.19
Polystyrene		0.20	98.67	0.68	0.45
Polyurethene		0.20	87.12	8.30	4.38
Polyvinyl Chloride		0.20	86.89	10.85	2.06
Linoleum		2.10	64.50	6.60	26.80
Rags		10.00	84.34	3.46	2.20
Vacuum Cleaner Dirt		5.47	55.68	8.51	30.34
Household Dirt		3.20	20.54	6.26	70.00
Municipal Wastes	3,000–6,000	15.35	37.65	0.6–15.0	15.0–27.0

the particulate pollutant problem, since small particles of this material can become entrained in the gaseous combustion products. These emissions are minimized in controlled-air incinerators by maintaining *nonturbulent* air conditions within the primary chamber. The proximate analysis of a waste is useful in estimating the conditions to be expected within a controlled-air incinerator. In general, the four major components of waste (volatile matter, fixed carbon, moisture, and noncombustibles) when incinerated have the effects described below.

1.3.1 Volatile Matter

The higher the volatile content of the waste, the greater the percent of total combustion occurring in the secondary chamber. This produces secondary chamber temperatures which are higher than primary chamber temperatures and reduces auxiliary fuel requirements.

Figure 1.3 is a representative temperature profile in both chambers for highly volatile waste controlled air incineration. The temperature of the secondary chamber stays well above that of the primary chamber while waste is fed to the unit. When the waste feed stops, the unit enters a burndown condition, during which a temperature inversion normally occurs—i.e., temperature in the primary chamber increases while the remaining fixed carbon is burned out. At the same time, the absence of volatiles in the secondary chamber reduces its temperature.

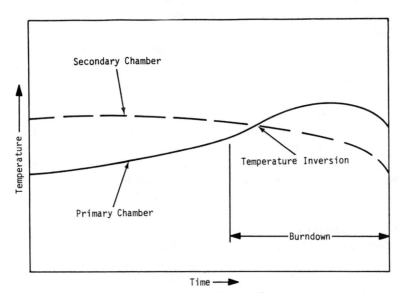

FIGURE 1.3 Temperature profile, high volatile waste content.

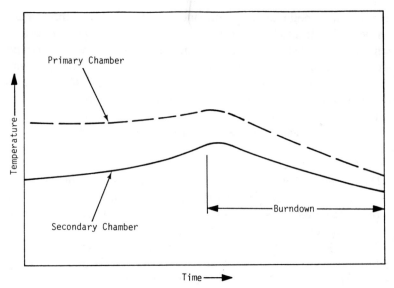

FIGURE 1.4 Temperature profile high fixed carbon waste.

1.3.2 Fixed Carbon

Increased fixed carbon content has the opposite effect of high volatiles—the higher the fixed carbon content, the greater the percent of total combustion occurring in the primary chamber. This produces primary chamber temperatures which are higher than the secondary, and may increase auxiliary fuel requirements in the secondary. Figure 1.4 shows a representative temperature profile for high fixed carbon controlled air incineration.

1.3.3 Moisture

Moisture in the waste must be evaporated before combustion can occur. Thus, the higher the moisture content, the greater the heat required to produce combustion. This increases the amount of auxiliary heat required in the primary chamber, and may also increase secondary chamber auxiliary heat requirements.

1.3.4 Noncombustibles

Noncombustibles have two primary effects: (1) they absorb heat, thus increasing the auxiliary heat requirement, and (2) they don't burn, which means they will end up in the effluent gas as ash and particulates unless measures are taken to prevent it.

1.3.5 Classification of Wastes

All wastes are categorized into types as shown in Table 1.3a. Noted are the basic 6 types plus the hazardous, toxic and 2 special types. Table 1.3b shows how typical wastes differ.

1.4 COMBUSTION

Combustion is a rapid chemical oxidation process of a fuel. In contrast, rusting of metals would be an example of extremely slow oxidation. Usually we desire complete combustion of the fuels to obtain the reaction products of full oxides and water. This is an exothermic reaction. More heat is available if the water product is condensed to liquid form—the higher heat of combustion. If the water is gaseous the heat release in lower—the lower heat of combustion. Partial oxide products, such as carbon monoxide, result in *much* lower heat releases. In addition to heat, energy in the form of light is also released. This is useful in incineration operations for indicating temperature.

To obtain complete combustion, the combustor must be properly designed to provide effective contact between the fuel and oxidant. In addition to this, it is necessary to have the 3 T's of combustion: time, temperature and turbulence. Time is necessary to permit the chemical reaction to proceed to completion after the reactants are together. Temperature is required to permit the reaction to proceed fast enough so that it will evolve enough heat to be self sustaining. Turbulence is necessary to mix and sustain intimate contacting of the reactants.

1.4.1 Types of Combustion

Combustion fuels can be either solids, liquids, gases or a combination of these. The oxidants may also be either in solid, liquid or gaseous form or in combination. Table 1.4 outlines the common basic combustion arrangements. There are many handbooks that describe combustion of gaseous and liquid fuels with air in great detail but few deal with combustion of solid fuels. Therefore, most of the following discussions relate to solid fuels with air, i.e. diffusion flames.

1.4.2 Theory of Combustion

The combustion of solids in air consists of four steps:
(1) When the solid fuel particles enter the furnace, they are rapidly heated and, as their temperature rises, moisture and volatile matter are

TABLE 1.3a
Classification of wastes to be incinerated.

Classification of Wastes		Principal Components	Approximate Composition (% by Weight)	Moisture Content (%)	Incombustible Solids (%)	BTU Value/Lb. of Refuse as Fired
Type	Description					
Class 0	Trash	Highly combustible waste. Paper, wood, cardboard cartons and up to 10% treated papers, plastic or rubber scraps: commercial and industrial sources	Trash 100%	10%	5%	8500
Class 1	Rubbish	Combustible waste, paper, cartons, rags, wood scraps, combustible floor sweepings: domestic, commercial and industrial sources	Garbage 20%	25%	10%	6500
Class 2	Refuse	Rubbish and garbage: residential sources.	Rubbish 50% Garbage 50%	50%	7%	4800
Class 3	Garbage	Animal and vegetable wastes: restaurants, hotels, markets, institutional, commercial, and club sources.	Garbage 65% Rubbish 35%	70%	5%	2500
Class 4	Animal solids and organics	Carcasses, organs, solid organic wastes: hospital, laboratory, abattoirs, animal pounds and similar sources.	Animal and human tissue 100%	85%	5%	1000

(continued)

TABLE 1.3a
(continued)

Class		Description			
Class 5	Gaseous, liquid or semi-liquid (non-hazardous)	Industrial process wastes incinerated directly through burner.	Variable	Dependent on major components	Variable	Variable 1,000 to 20,000 Btu/lb.
Class 6	Semi-solid and solid (non-hazardous)	Combustibles requiring rotary retort equipment.	Variable	Dependent on major components	Variable	Variable
Class—	Hazardous (as defined by RCRA*)	Chemical, petroleum, and pharmaceutical wastes; herbicides; pesticides; explosives; plastics; etc.	Variable	Dependent on major components	Variable	Variable
Class—	Toxic (as defined by TSCA**)	PCB's etc.	Variable	Dependent on major components	Variable	Variable
Class—	Metal Recovery	Gold, silver, cooper, aluminum, etc.	Variable	Dependent on major components	Variable	Variable
Class—	Filter media reclamation	Diatomaceous earth, carbon, polyols, etc.	Variable	Dependent on major components	Variable	Variable

Courtesy of C. E. Raymond, Form 881-5M-C1 [3].
*RCRA—Resource Conservation and Recovery Act of 1976.
**TSCA—Toxic Substances Control Act of 1978.

TABLE 1.3b
Typical waste variations.

Description	% Moisture	% Combustible	% Ash	Btu/lb
Average Wood	0	100	0	9000
Average Wood	10	87	3	7800
Average Wood	30	67	3	6000
Average Wood	50	47	3	4200
Average Paper	0	100	0	7700
Average Paper	10	89	1	6850
Average Paper	10	86	4	6600
Average Paper	20	79	1	6100
Average Paper	20	76	4	5900
Municipal Waste	25	50	25	4500
Polyethylene	0	100	0	20,000

driven off. The volatile matter ignites and burns with a hot luminous flame. After the volatile matter has been burned, the residue of fixed carbon starts to burn without visible flame and combustion proceeds to completion.

(2) The particles attain a temperature of about 2000 °F before either ignition or a significant amount of pyrolysis occurs. Ignition precedes rapid devolatilization and originates on the surfaces of solid particles instead of in gaseous volatiles.

(3) According to the modes of ignition, solid fuels can be classified as: (a) fuels with surface ignition only, (b) with gas phase ignition only, and (c) fuels with hybrid behavior where a Transition Ignition Phase exists.

(4) The ignitability of solid fuels is determined not by its total volatile content, but rather by the surface properties of both the original and partially charred or carbonized particles.

TABLE 1.4
Types of combustion.

Initial State		Fuel and Oxidant	
Fuel	Oxidant	Mixed	Separate
Gas	Air	Premixed flame	Diffusion flame
Liquid	Air	Premixed flame	Diffusion flame
Liquid	Liquid	Monopropellant combustion	—
Solid	Air	—	Diffusion flame
Solid	Solid	Propellant combustion, explosion	—

TABLE 1.5
Chronology of the combustion process.

```
              Heating and Minor Devolatilization
              ▓▓▓▓▓▓▓▓▓▓▓
                        Ignition
                        ▓
                        Major Devolatilization
                        ▓▓▓▓▓▓▓
                              Burning of the Carbon Residue
                              ▓▓▓▓▓▓▓▓▓▓▓▓▓▓▓▓▓▓▓▓▓▓▓▓▓▓
 0 ─────────────────────────────────────────────────────────
                              Time
```

Currently there are two proven methods for firing solid fuels in industrial boilers, stoker or pulverized fuel firing. The chronology of the fundamental combustion process as the fuel enters the furnace is basically the same for either firing system. In all systems the 3 T's of combustion are required. Time, Temperature and Turbulence. The chronology of the combustion process can be envisioned as shown in Table 1.5.

Combustion is a chemical reaction and all reactions are reversible to some extent; no reaction ever goes to absolute completion. A stoichiometric combustion equation has just the amounts of reactants required by the balanced chemical equation for theoretical complete combustion. A reaction can be made more complete by increasing the concentration of the reactants, so in combustion operations it is common to use more than the theoretical stoichiometric amount of oxygen. Combustion of organic fuels having a general generalized formula C_mH_n have the stoichiometric combustion equation with oxygen represented as

$$C_mH_n + (m + \frac{n}{4} O_2) \rightarrow m\ CO_2 + (n/2)\ H_2O$$

Using as an example methane (CH_4), the simplest hydrocarbon, the stoichiometric combustion equation would be:

$$CH_4 + 2[O_2 + (\frac{79}{21})\ N_2] \rightarrow CO_2 + 2H_2O + 2\,(\frac{79}{21})\ N_2$$

| Fuel | Oxidant Air | Diluent | Combustion Products | Diluent |

1.4.3 Excess Air

Excess (X's) air is necessary for complete combustion. The percent excess air can be determined by any of the following three equivalent relationships.

$$\% \ X\text{'s air} = \frac{Total\ Air - Theoretical\ Air}{Theoretical\ Air} (100)$$

$$= \frac{Excess\ Air}{Theoretical\ Air} (100)$$

$$= \frac{Excess\ Air}{Total\ Air - Excess\ Air} (100)$$

Theoretical air is that amount required to stoichiometrically convert all combustible species (mainly C, H and S) to complete normal products of combustion (i.e., CO_2, H_2O and SO_2). These relationships are stated as mole ratios of air which equal volume ratios. Although moles of air are specified in each term, the moles of oxygen can be used instead if applied consistently. Thus, it is proper to call this relationship either "excess air" or "excess oxygen." However, excess oxygen is also the percentage of unburned oxygen in the flue gases and can lead to confusion of terms if care is not exercised.

The generalized combustion equation with excess air is:

$$CH_4 + (2 + a)\ [O_2 + (\tfrac{79}{21})\ N_2] \rightarrow CO_2 + 2H_2O + (2 + a)\ (\tfrac{79}{21})\ N_2 + aO_2$$

| Fuel | Air | Combustion Products | Diluents |

1.4.4 Air-Fuel Ratio

Air-fuel ratio (A/F) is the mass of air per unit mass of fuel. This can be obtained directly from the equation. For example, using the theoretical combustion of CH_4 with stoichiometric air, the air-fuel ratio is

$$A/F = \frac{(2\ mole\ O_2)\ (\frac{100\ moles\ air}{21\ moles\ O_2})\ (\frac{29\ g\ air}{g\ mole})}{(1\ mole\ CH_4)\ (\frac{16\ g}{g\ mole})} = 17.3$$

As the fuel becomes heavier, i.e. the ratio of moles of hydrogen per mole of carbon decreases. For example, at stoichiometric conditions gasoline ($\sim C_8H_{18}$) has an A/F of 15 and for pure carbon it is 11.5. The fuel/air ratio (F/A) is obviously the reciprocal of the A/F ratio.

1.4.5 Equivalence Ratio

Often it is desirable to compare the richness or leanness of combustion for different solid fuels. The equivalence ratio, ϕ, is convenient for this type of comparison, and it may be defined as the quotient of the actual fuel-air ratio divided by the stoichiometric fuel-air ratio.

$$\phi_{FA} = \frac{(F/A) \ actual}{(F/A) \ stoichiometric \ (theoretical)}$$

or, alternatively, in terms of air-fuel ratios

$$\phi_{FA} = \frac{(A/F) \ actual}{(A/F) \ theoretical}$$

For a stoichiometric mixture $\phi = 1.0$. However, it should be recognized that even though the equivalence ratio defined in the equations above have identical values at stoichiometric, they are not identical for "off-stoichiometric" (rich or lean) mixtures because (F/A) and (A/F) ratios are reciprocals. This is illustrated in Table 1.6.

1.4.6 Factors that Affect the Combustion Process

For solid fuels there are several important criteria that must be considered. These are listed

(1) Solid fuel

1. Nature of the fuel-volatile content, type of char, agglomerating tendency, mineral matter, impurities
2. Particle Size
3. Temperature
4. Oxygen Concentration
5. Mixing and Recirculation
6. Flame Size

(2) Ignition

1. Flame Stability
2. Desire rapid ignition so that more time will be available for burnout.

16 CONTROLLED AIR INCINERATION

TABLE 1.6
Equivalence ratio.

Mixture	Value
Rich	$\phi_{AF} > 1$
Stoichiometric	$\phi_{AF} = 1$
Lean	$\phi_{AF} < 1$
Rich	$\phi_{FA} < 1$
Stoichiometric	$\phi_{FA} = 1$
Lean	$\phi_{FA} > 1$

1.4.7 Devolatilization characteristics
There are four factors to consider

(1) Rapid reaction—requires about 10% of the combustion time.
(2) Volumetric (rather than a surface) reaction for pulverized firing.
(3) Rate of devolatilization is directly proportional to the volatile content remaining in the fuel (first order reaction) and increases exponentially with temperature.
(4) Most of the volatiles (2/3) burn homogeneously—remainder burns with the char.

Since the agglomerating property of the fuels is the result of the particles transforming into a plastic or semi-liquid state on heating, it reflects a change in the surface area of the particle. This surface change is manifested by a transformation of the particle from an angular, irregular shape into a spherical or sphere-like particle. Also, the surface character of the particle changes from a porous, irregular absorptive surface to a glass-like non-porous surface. Thus, with the application of heat, agglomerating coals would tend to develop a non-porous surface while the surface of non-agglomerating coals would become even more porous with pyrolysis.

1.4.8 Combustion Simplified
Organic waste generally consists of carbon (C), hydrogen (H) and oxygen (O), with smaller amounts of sulfur (S), nitrogen (N) and chlorine (Cl). For purposes of this discussion, the S, fuel bound N and the Cl will be ignored.
Combustion occurs according to the following chemical equations:

$$C + O_2 \rightarrow CO_2$$

$$2H_2 + O_2 \rightarrow 2H_2O$$

These equations state that carbon and hydrogen combine with oxygen to form carbon dioxide and water vapor respectively. Oxygen in the waste passes through the combustion process unchanged. Heat is liberated during combustion in the amount of 14,100 BTU per pound of carbon burned and 61,000 BTU per pound of hydrogen burned. The amount of oxygen required by the equation above is called "stoichiometric" or theoretical oxygen. This means that the exact amount of oxygen required for the combustion of carbon and hydrogen is involved. Because of inadequate mixing in commercial equipment, and temperature restrictions on the refractory, it is common practice to supply more oxygen (or combustion air) than is theoretically required so as to ensure that all carbon and hydrogen will be completely burned. This additional oxygen is called "excess air." Maximum combustion temperatures are produced at the stoichiometric condition. As the amount of air is increased or decreased from the stoichiometric point, the combustion temperature is lowered. Figure 1.5 shows the combustion temperature as a function of air supply. The basic approach to controlled-air incineration is to maintain primary chamber conditions to the left of the stoichiometric point, and secondary combustion conditions to the right

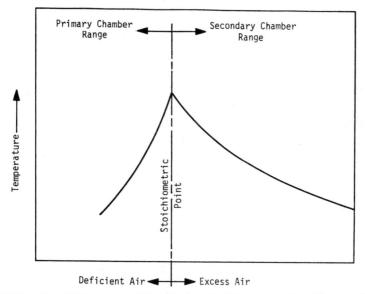

FIGURE 1.5 *Air and temperature relationships in controlled-air incinerators. (Courtesy R. E. McRee [4])*

18 CONTROLLED AIR INCINERATION

of the stoichiometric point. The two most significant factors in controlled air incineration are excess air and waste composition. Excess air directly or indirectly affects virtually all of the operating parameters of the incinerator—chamber temperatures, heat recovery efficiency, auxiliary fuel requirements, and ash and particulate emissions, among others.

In general, by reducing the air supply to the unit increases combustion retention time, and reduces the amount of auxiliary heat required to achieve adequate combustion temperatures. Turbulence is produced through combustion chamber design, by use of an over-fired air supply in the top of the primary chamber, and/or by the addition of combustion air to the secondary chamber.

On the other hand, the chemical composition of the waste determines the amount of air required for complete combustion, which, in turn, establishes the point at which additional air becomes "excess air."

1.4.9 Flue Gas Composition

An example can be given for the combustion of hydrocarbons. Assume a properly designed and operated combustion system. The theoretical stoichiometric complete combustion equation of the simplest hydrocarbon (natural gas, methane) with air is

$$CH_4 + 2\ O_2 + 7.52\ N_2 \rightarrow CO_2 + 2H_2O + 7.52\ N_2$$

Air is considered to be 79% N_2 and 21% O_2. Mole fraction equals volume fraction so the moles of N_2 equals (2 moles O_2) (79 moles N_2/21 moles O_2) or 7.52 moles. The presence of nitrogen and oxygen at elevated temperatures results in the formation of nitric oxide (NO) which converts to nitrogen dioxide (NO_2) in the atmosphere. The amount of NO is very small compared to the SO_2, H_2O and N_2. The composition of the flue gas produced by the equation above if all components are gases, can be determined using the gas laws:

$$\frac{1\ mole\ CO_2}{(1 + 2 + 7.52)\ mole\ total}(100) = (\frac{1}{10.52})(100) = 9.5\ mole$$

or volume % CO_2

$$(\frac{2}{10.52})(100) = 19.0\%\ H_2O$$

$$(\frac{7.52}{10.52})(100) = 71.5\%\ N_2$$

This is called a wet gas analysis. The molecular weight of this flue gas is then 27.6 lbs/lb mole and the density at SC is

$$\left(\frac{27.6 lb}{359 ft^3}\right)\left(\frac{460 + 32}{460 + 70}\right) = 7.14 \times 10^{-2} \, lb/ft^3$$

Continuing the methane combustion example, 5% excess air would be 105% theoretical air and the resultant balanced chemical equation would show (1.05)(2) moles of oxygen and (1.05)(7.52) moles of nitrogen

$$CH_4 + 2.1 \, O_2 + 7.9 \, N_2 \rightarrow CO_2 + 2 \, H_2O + 0.1 \, O_2 + 7.9 \, N_2$$

The complete combustion example shows, as before, no CH_4 or CO in the products but the excess O_2 now is present. Theoretical composition of the combustion gas is 9.09% CO_2, 18.18% H_2O, 0.91% O_2 and 71.82% N_2. Dry gas combustion is:

$$\left(\frac{9.09}{100 - 18.18}\right)(100) = \left(\frac{9.09}{81.82}\right)(100) = 11.11\% \, CO_2$$

$$\left(\frac{0.91}{81.82}\right)(100) = 1.11\% \, O_2$$

$$\left(\frac{71.82}{81.82}\right)(100) = 87.78\% \, N_2$$

This type of analysis is obtained directly by measuring the actual flue gas composition with an Orsat gas analyzer.

This example can be expanded for other values of theoretical air. Figure 1.6 is produced in this manner. Note that at less than 100% air, any or all of CH_4, CO and C must be present. Particulate matter as well as gases can be present; however, this is not significant for CH_4 combustion as only near and greater than 100% theoretical air is being considered. Nitrogen is the largest component of typical flue gases and is the major factor responsible for the large size of pollution control equipment. It also accounts for most of the gas pumping costs and for the greatest energy (heat) loss. Use of oxygen instead of air for combustion would reduce these wastes but the cost and energy of producing oxygen must be balanced against the savings and the losses in turbulence and dilution.

The effects of excess air on % CO_2 and % moisture in the flue gas are

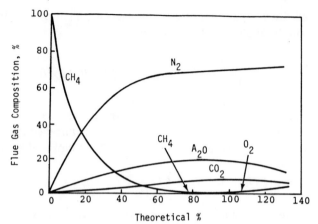

FIGURE 1.6 Theoretical combustion of natural gas with air.

shown in Table 1.7 for various hydrogen to carbon mass ratios (H/C). The water values are those from fuel hydrogen only.

1.4.10 Flue Gas Quantity

Volume of flue gas can be related to the fuel consumption rate. If the combustion equation is known this can be used. For example, using the

TABLE 1.7
Effects of excess air on CO_2 and moisture.

Fuel	Approximate H/C ratio	% Excess Air	% by Volume in Flue Gas	
			CO_2	H_2O
Gaseous	0.31	0	12.0	22
		20	9.8	18
		40	8.3	16
		80	6.3	12
		100	5.7	11
Liquid	0.12	0	15.4	8
		20	12.8	7
		40	10.9	6
		80	8.4	4
		100	7.5	3
Solid	0.05	0	18.4	4
		20	15.3	3
		40	13.0	3
		80	10.1	2
		100	9.0	2

equation with 5% excess air, the volume of wet flue gas in actual cubic meters at 150 C per mole of CH_4 is

$$\left(\frac{1 + 2 + 0.1 + 7.9 \text{ moles total gas}}{\text{mole } CH_4}\right)\left(\frac{22.4 \text{ l}}{\text{g mole}}\right)\left(\frac{273 + 150 \text{ K}}{273 \text{ K}}\right)\left(\frac{m^3}{1,000 \text{ l}}\right)$$

$$= 0.382 \text{ Am}^3/\text{mole } CH_4$$

This may also be reported as the volume of dry gases at standard conditions by simply subtracting the moles of water vapor in the product gases and proceeding from there. More likely, only the flue gas analysis is known. Using the analysis, the volume of dry flue gas at SC per mole CH_4 is:

$$\left(\frac{100 - 18.18 \text{ moles dry gas}}{9.09 \text{ moles } CH_4}\right)\left(\frac{359 \text{ ft}^3}{\text{lb mole}}\right)\left(\frac{460 + 32}{460 + 70}\right)$$

$$= 3000 \text{ std. ft.}^3/\text{mole } CH_4$$

Emissions, expressed in terms of equivalence ratios, for combustion of hydrocarbon fuels can be approximated from Figure 1.7.

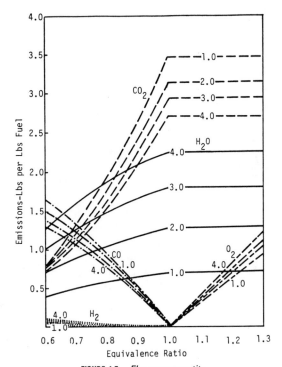

FIGURE 1.7 Flue gas quantity.

TABLE 1.8
U.S. EPA survey of incinerator types [5].

No.	Company	Type Incinerator* Built	Units in H.W. Service in U.S.	Unit* Capacity	Nature of Wastes	Comments
1	Air Resources	L.I. Followed by Catalytic Converter	1	5 gal/h	Liquid	Firm is primarily in fume incin. business. Supplies H.W. incins. only on request
2	Basic Environmental Engineering, Inc.	Pulse Hearth	2	6×10^6 Btu/h	Paint Residues	Does not solicit business
				20×10^6 Btu/h	residual cutting oils, other	
3	Baumco, Inc.	L.I.	0			
4	Bayco Industries of California	Fixed hearth, Multichamber Afterburner	≤12	Mostly 200–300lb/h Up to 600 lb/h	Solids, sludges, liquids	Primarily a burner vendor
5	Bigelow-Liptak Corp.	L.I. R.K.	19- - - 1	Mostly 7–10×10^6 Btu/h or 30–40×10^6 Btu/h	Mostly chlorinated HC	

(continued)

TABLE 1.8 (continued)

No.	Company	Type Incinerator* Built	Units in H.W. Service in U.S.	Unit* Capacity	Nature of Wastes	Comments
6	Brule' C.E. & E., Inc.	L.I.	5	$3.5–8.5 \times 10^6$ Btu/h	Chlorinated HC	Vendor also mfgs. fume incinerators
7	Burn-Zol Corp.	Fixed Hearth	18	3–100 gal/h on liquids 200–1,800 lb/h on solids or solids & liquids	Plating wastes, Paint solvents and sludges, Plastic residues	Vendor also mfgs. fume incinerators
8	C&H Combustion	R.K.	2	10×10^6 Btu 1,000–2,000 lb/h	Pharmaceutical wastes	
				$120 \& 10^6$ Btu/h	Commercial industrial wastes	
9	CEA Combustion, Inc.	L.I. L.I. & R.K.	1 ~17 L.I. ~ 3 R.K.	? 10×10^6 or 30–40 $\times 10^6$ Btu/h	Organics, silicones Liquids, Solids	Also mfgs. fume incinerators. Sees sales trend to R.K.

(continued)

TABLE 1.8 (continued)

No.	Company	Type Incinerator* Built	Units in H.W. Service in U.S.	Unit* Capacity	Nature of Wastes	Comments
10	CE Raymond, Bartlett-Snow Division	R.K., L.I., R.B.	16 R.K. (+3 in const.) 0 L.I. 0 F.B.	5–7 × 10^6 Btu/h-6 units 10 × 10^6-1 unit 25 × 10^6-1 unit 50 × 10^6-8 units	Solids, Semisolids, Liquids	A major supplier of R.K.
11	CICO, Inc.	"Semisuspension" Process	0	---	---	Test unit in development
12	Coen Co.	L.I.	2; others possibly	75 × 10^6 and 67 × 10^6 Btu/h afterburner for Cin. Municipal L/Fl ——————— 20,000 gal/d unit in construction	Mixed H.W. and Industrial Wastes	

(continued)

TABLE 1.8 *(continued)*

No.	Company	Type Incinerator* Built	Units in H.W. Service in U.S.	Unit* Capacity	Nature of Wastes	Comments
13	Commercial Fabrication and Machine Co., Inc.	"Novel-Type Package Incinerator"	0	200 lb/h		Unit in development. Allegedly usable for H.W.
14	Copeland Associates, Inc.	F.B.	2	$8-9 \times 10^6$ Btu/h	Refinery Wastes	
15	Dorr Oliver, Inc.	F.B.	5	35×10^6 Btu/h 50×10^6 Btu/h $2@67 \times 10^6$ Btu/h	Pulp Mill Black Liquor Neutral Sulfite Liquor	
				370 ton/d	Wet Sludges Contg. Oily Matter	
16	Dravo Engineers and Constructors	R.K., L.I.	0	---	Multiple types of waste	Designs and installs complete H.W. facilities based on Bayer/Steinmuller technology

(continued)

TABLE 1.8 (continued)

No.	Company	Type Incinerator* Built	Units in H.W. Service in U.S.	Unit* Capacity	Nature of Wastes	Comments
17	Ecologenics Corp.	Salt Bath	0	---	---	Unit in Development
18	Econo-Therm Energy Systems Corp.	Fixed Hearth, w/Starved Air	≤10	50 to 2,500 lb/h	N.A.	Also builds cofiring municipal incins. R&D on FB unit
19	Enercon Systems, Inc.	Fixed Hearth	2; others possibly	600 lb/h 240 ton/d	Insecticides Munic. wastes plus liquid H.W.	Also developing a rotary hearth design
20	Energy, Inc.	F.B.	0	---	---	Pilot plant is test burning HW
21	Entech Industrial Systems, Inc.	L.I. and Multiple Hearth	6 L.I. --- 6 M.H.	50-500 gal/d ------ 1,000-1,500 lb/d	liquids ----- solids	Also mfgrs. fume incinerators
22	Environmental Control Products, Inc.	Fixed, Stepped Hearth	Unkn.	300-1,000 lb/d		Also builds incin. for radioactive wastes and for municipal wastes

(continued)

TABLE 1.8
(continued)

No.	Company	Type Incinerator* Built	Units in H.W. Service in U.S.	Unit* Capacity	Nature of Wastes	Comments
23	Environmental Elements Corp.	R.K.	0	---	Liquids, solids and fumes	Marketing European technology that is widely used in Europe and Japan
24	Fuller Co.	F.B. & R.K.	Unkn.	---	---	Has F.B. and R.K. units in operation but nature of wastes is unknown
25	HPD, Inc.	L.I.	0	---	Unit specific for salt-containing organic/aqueous wastes	Unique European technology
26	Hirt Combustion Engineers	L.I.	5-10 (assume 8)	1-10 gal/min	Haz. liquids	All units custom built. Also engineers fume incinerators.

(continued)

TABLE 1.8 (continued)

No.	Company	Type Incinerator* Built	Units in H.W. Service in U.S.	Unit* Capacity	Nature of Wastes	Comments
27	Howe-Baker Engineers, Inc.	L.I., R.K.	0	---	---	Company is quoting on several L.I. facilities. Company also manufactures burners.
28	Industronics, Inc.	R.K.	0	---	Liquids and solids	Designs emphasize energy and precious metals recovery. Actively marketing.
29	International Incinerators, Inc.	R.K.	3 +2 in const.	$5\text{--}100 \times 10^6$ Btu/h	Chlor. H.C. Tars	
30	John Zink Co., The	L.I. (Primarily) Oxidative or Starved Air	75(±10) (since Jan. 1969)	$5\text{--}300 \times 10^6$ Btu/h Avg. is $30\text{--}50 \times 10^6$ Btu/h	All types liquids	75% of units sold in last 3 yrs. included heat recovery. Recently, this is 95%. Also mfg. fume incinerators

(continued)

TABLE 1.8
(continued)

No.	Company	Type Incinerator* Built	Units in H.W. Service in U.S.	Unit* Capacity	Nature of Wastes	Comments
31	Kelley Co., Inc.	F.H. Starved Air, for Solids and/or Liquids	3 in const.	1 @ 9 × 10⁶ Btu/h 2 @ 3 × 10⁶ Btu/h	Waste oils, solvents, aqueous solns, contg. organics, solids	
32	Lurgi Corp. (formerly BSP Envirotech)	F.H. Multiple Hearth, R.K., F.B.	5 (3 F.H. 1 Multihearth 1 R.K.)—— 0 F.B.	1,200 lb/h, some liquids	---	Designs and installs (turnkey) complete facilities
33	McGill, Inc.	L.I.	0	---	---	Now quoting on several units for H.W. service
34	Met-Pro Corporation Systems Div.	R.K., L.I.	2 R.K. 1 L.I.	50 gal/h 120 gal/h		
35	Midland-Ross Corp.	Rotary Hearth Pyrolyzer	1 (2nd near startup) (assume 2)	3,600 lb/h 5,800 lb/h	Organic solids, sludges	Unit touted for use on sludges with large metal content

(continued)

TABLE 1.8
(continued)

No.	Company	Type Incinerator* Built	Units in H.W. Service in U.S.	Unit* Capacity	Nature of Wastes	Comments
36	Morse Boulger	Fixed Hearth	3	650–1,000 lb/h	Unkn.	Relatively new in this business Actively marketing
37	Niro Atomizer Inc.	F.B.	0	—	—	
38	P&T Manufacturing Co.	F.H. Starved Air, R.K.	5 F.H. 0 R.K.	25–1,000 lb/h	Paint solvents Chemical slurries	
39	Peabody International Corp.	L.I. Single or Dual Chamber	2; others possibly	99×10^6 Btu/h 41×10^6 Btu/h	Oil and acid sludge Range of solvents, paints, tars, oils, etc.	Also mfgs fume incinerators, burners, and other related equipment
40	Perstorp, Inc.	Oxidative Incin. of H.W. Contained in Open-End Drums	0	—	Liquids and solids	One H.W. facility operating in Sweden
41	Plibrico Co.	F.H.	Unkn.			Custom-engineered units; not actively marketing

(continued)

TABLE 1.8
(continued)

No.	Company	Type Incinerator* Built	Units in H.W. Service in U.S.	Unit* Capacity	Nature of Wastes	Comments
42	Prenco, Inc.	L.I.	~22	3–112 × 10^6 Btu/h	Varied	Constructing a mobile test unit. New company.
43	Pyro Magnetics Corp.	Pyrolysis w/ Induction Heating	0	---	---	Actively marketing; making quotations in several units
44	Rockwell International	Molten Salt Bath	0	---	---	
45	Shirco, Inc.	L.I., Infrared, Belt-Driven Chamber w/ or w/o Gas-Fired Afterburner	1 Infrared unit in constr. 0 L.I.	50 lb/h	Phenolic sludges	Infrared pilot unit available for testing
46	Sunbeam Equipment Corp., Comtro Div.	Controlled Air 2-Stage, F.H.	2; Others Possibly	200 & 800 lb/h	Solids, plastics wastes	Not highly interested in H.W. market

(continued)

TABLE 1.8
(continued)

No.	Company	Type Incinerator* Built	Units in H.W. Service in U.S.	Unit* Capacity	Nature of Wastes	Comments
47	Sur-Lite Corp.	L.I.	3	60, 200, 500 gal/h	Oily wastes shale oil residues	Has proprietary scrubber design
48	TR Systems	R.K.	8	4 @ 1-3 × 10^6 Btu/h 1 @ 7 × 10^6 Btu/h 1 @ 10 × 10^6 Btu/h 1 @ 20 × 10^6 Btu/h 1 @ 28 × 10^6 Btu/h	Unkn.	
49	Tailor and Co., Inc.	L.I., F.B.	2 F.B. --- 0 L.I.	2 and 10× 10^6 Btu/h	Unkn.	75 × 10^6 Btu/h; F.B. unit in design
50	Thermal Processes, Inc.	F.B.	0	---	---	
51	ThermAll, Inc.	R.K.	0	---	---	One unit in construction
52	Trane Thermal Co.	L.I.	54	5 to 130 × 10^6 Btu/h	Chlor. H.C. tarry still bottoms, formaldehyde, etc.	Also mfg. fume incinerators

(continued)

TABLE 1.8
(continued)

No.	Company	Type Incinerator* Built	Units in H.W. Service in U.S.	Unit* Capacity	Nature of Wastes	Comments
53	Trofe, Inc.	Starved Air, Oscillating, R.K.	0	---		25×10^6 Btu/h; Test unit for liquids, solids and mixtures
54	The United Corp.	L.I.	3	One 10 & two 15×10^6 Btu/h		6×10^6 Btu/h; Test unit available
55	U.S. Smelting Furnace Co.	F.H.	1	4.6×10^6 Btu/h	Solvents	Does not seek H.W. incin. business
56	Vulcon Iron Works, Inc.	R.K.	1	25 ton/d	Solids and liquids	
57	Washburn and Granger Co., Inc.	Reciprocating Grate; Controlled Air	1	NA	Liquids and fumes	

*L.I. = liquid injection, R.K. = rotary kiln, F.B. = fluidized bed, F.H. = fixed hearth, H.W. = hazardous waste, N.A. = not available, SI Conversion: L = gal × 3.79; kg = lb × 0.454; kJ = Btu × 1.055; w = Btu/h × 0.293.

1.5 INCINERATOR TYPES

The U.S. EPA conducted a survey [4] thru MITRE Corporation in 1981 with 119 vendors of incinerator systems. A summary of results from these government data are given in Table 1.8. The fixed hearth type represent about 17.3% of all hazardous waste incinerators of which over 340 incinerators (total), manufactured by 29 companies, have been put into hazardous waste service since 1969.

REFERENCES

1 Danielson, John A. (ed.), "Air Pollution Engineering Manual," U.S. EPA, OAQPS (May 1973).
2 McIlvaine, Robert (ed), "The Electrostatic Precipitator Manual," Chapter 9, page 502.2 (1976).
3 Raymond, C. E., Form 881-5M-Cl, "I.I.A. Incinerator Standards," Incinerator Institute of American, page 1968-5A (November 1968).
4 McRee, Robert E., "Operation and Maintenance of Controlled-Air Incenerators," 76th Annual APCA Meeting, paper No. 83-59.1, Atlanta (June 1983).
5 Frankel, I., et al, "Hazardous Waste Incineration Facility Data Base," U.S. EPA IERL, Cincinnati, OH, Contract No. 68-03-3021 (1983).

CHAPTER 2

Incinerator Operation

2.1 TEMPERATURE CONTROL AND EXCESS AIR

The temperature rating of the refractory is one of the most important factors affecting the operation of the incinerator. Most controlled air incinerators use a nominal 2800–3000 °F. refractory. However, in practice, it is desirable to limit the temperature in the secondary chamber to about 2200 °F (assuming a 2800–3000° nominal refractory). Excess air is used to control the temperature. Figure 2.1 shows that the conditions described above require approximately 50% excess air, in the case of Type I waste with 25% moisture [1].

Generally, the combustion air is supplied as approximately 20–25% underfire air in the primary, and the remaining air as combustion air in the secondary.

2.2 LOADING

The method used to load the waste into the incinerator depends on the waste itself, and on the scheduling of incinerator operations. If a single shift operation is utilized, the unit can be "overloaded" with waste, which is left to burn down during the "non-operational" period. (Such units were often called "stuff and burn" units.) The underfire air to the primary is controlled so as to produce only the volatiles which can be handled by the secondary. Figure 2.2 is a sketch of this type incinerator. The secondary can be designed to handle nearly any level of volatiles. It could, in fact, be merely an afterburner with air injection (which would, of course, limit the amount of waste "stuffed" into the primary).

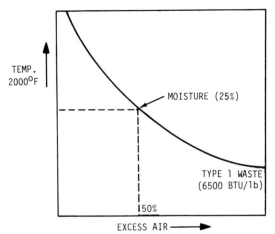

FIGURE 2.1 Regulation of incinerator temperature with excess air.

If more than one shift is required, the charging rate of the primary must be roughly equal to the burn rate. Thus, most controlled-air incinerators contain complex underfire air systems which maximize carbon oxidation while minimizing turbulence within the burning mass to reduce particulate entrainment. As a result of additional underfire air, most continuous-duty controlled-air units normally operate at higher primary chamber temperatures than units designed primarily for single shift operation.

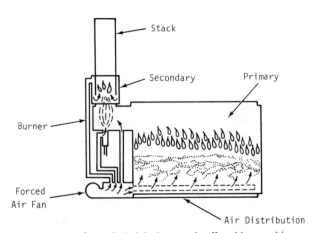

FIGURE 2.2 Controlled air incinerator (stuff and burn unit).

Incinerator Operation 37

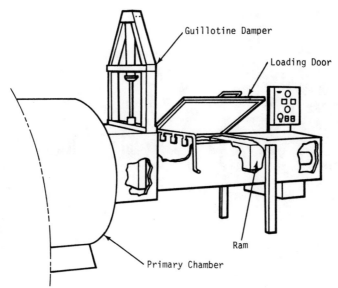

FIGURE 2.3a *Ram feed loading device for controlled air incinerator.*

A better loading system is the ram feed as shown in the Figure 2.3a schematic. Here, the waste is charged, the door is closed, and the guillotine is opened. Then the waste is pushed into the primary chamber by the ram. This provides better control of air and fuel. Smaller fuel charges can be made and the air leaks are minimized. Figures 2.3b through 2.3e are photographs of Comtro® hydraulic loading arrangements and incinerators.

The best is the continuous burn type of incinerator. Figure 2.4 is a schematic diagram of a modern continuous burn controlled air incinerator. It has been found in practice that it is impractical to burn 100% of the carbon in continuous burn incinerators. This is because the process of burning the final 10% or so of carbon in the ash results in excessive particulate emissions, which would be more difficult and expensive to control than can be justified by burning the carbon. (In the case of a waste heat boiler, the recoverable heat is reduced by the amount which would be recovered if the carbon were burned.) If a hazardous waste is being burned then it is important that all the undesirable constituents be completely burned out so that the ash/non-combustible can be disposed of without regulatory problems at a sanitary landfill.

® = Registered Trade Name.

FIGURE 2.3b 4 3/4 yd³ hydraulic loader for pathological waste—pan on wall controls loader and incinerator located outside. Courtesy John Zinc Co., Comtro Div.

FIGURE 2.3c 400 lb/hr pathological incinerator and stack with spark catcher—dual fuel fired (natural gas and #2 fuel oil); Comtro Model A-24. Loading ram is behind wall. Courtesy John Zink Co., Comtro Div.

FIGURE 2.3d Tipping floor with one yd³ hydraulic loader for Comptro A-24 incinerator. Courtesy John Zinc Co., Comptro Div.

FIGURE 2.3e One yd³ hydraulic loader, ram type, for Comptro A-24 incinerator. Courtesy John Zinc Co., Comptro Div.

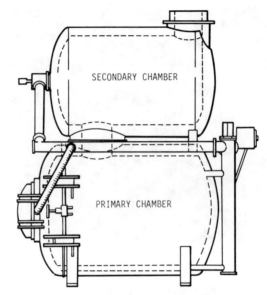

FIGURE 2.4 Modern style continuous burn controlled air incinerator.

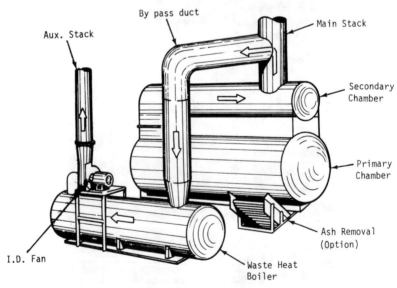

FIGURE 2.5 Controlled air incinerator with waste heat boiler.

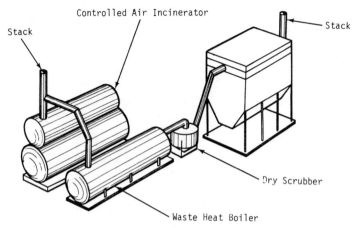
FIGURE 2.6   Controlled air incinerator with dry scrubber & baghouse.

Figures 2.5 and 2.6 show the physical layout and design of a typical controlled air incinerator with heat recovery and control devices.

2.3 SYSTEM BEHAVIOR

Figure 2.7 and 2.8 describe the system behavior typical standard multiple chamber and controlled air incinerator. These peaks occur during charging resulting from infiltration air. When the charging door is open, additional air causes a slight increase in primary chamber temperature as the result of localized burning of some volatile material in that chamber. This causes a slight temperature decrease in the secondary chamber. When the charging door closes, the primary chamber begins to cool as a result of operation at less than stoichiometric air, and the secondary temperature increases due to increased combustion of volatiles generated by the new waste charge.

2.4 AIR CONTROL

There are basically three methods of air control:

- Manual (operator) control

42 CONTROLLED AIR INCINERATION

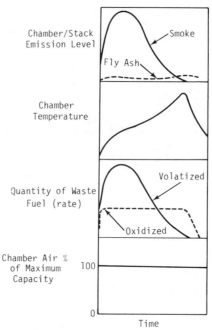

FIGURE 2.7 System behavior of standard multiple chamber incinerator batch-burning of high Btu waste.

- Oxygen modulated control
- Temperature modulated control

2.4.1 Manual Air Control

The least expensive method, manual air control is dependent on operator judgment and experience. It is rather unreliable as a method of controlling burn rate, refractory temperatures, particulate emissions and other operating variables.

2.4.2 Oxygen Modulated Air Control

This is a very complex and expensive method of controlling secondary combustion air, and is not widely used.

2.4.3 Temperature Modulated Air Control

Temperature modulated controls are designed to maintain a predetermined temperature level in the secondary combustion chamber. The controller is set for this temperature, and operates so as to admit increased combustion air if the temperature rises above the set point, and restrict

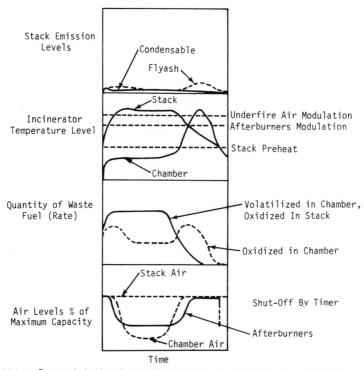

FIGURE 2.8 System behavior of controlled air incinerator batch-burning of high Btu waste.

air flow if the temperature falls below the setpoint. The effectiveness of these controls can be monitored by Orsat analysis of the effluent gases, testing the gas concentrations of carbon dioxide and/or oxygen.

Figures 2.9, 2.10 and 2.11 supplement Table 1.7 in showing the excess air effects on effluent gas concentrations for various examples of wastes and temperature control setpoints. For example, for municiple type waste and a secondary combustion chamber temperature of 2000°F, Figure 2.9 indicates that approximately 60% excess air would be required. At this level, Figure 2.10 shows the effluent can be expected to contain approximately 7.5% oxygen and Figure 2.11 shows 10.5% carbon dioxide.

Controlled air incinerators generally have a second temperature controller for the secondary auxiliary burner. This controller has low and high setpoints which serve to preheat the combustion chamber. The auxiliary burner preheats the chamber to the low set point, at which time, waste can be fed to the unit. The auxiliary burner continues to fire until the high setpoint is reached (in the range of 1800°F to 2000°F, depending

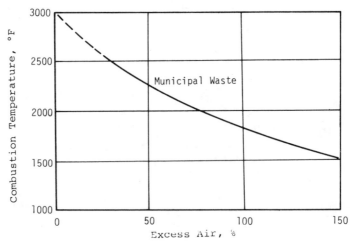

FIGURE 2.9 Typical combustion temperature vs % excess air for municipal waste incineration.

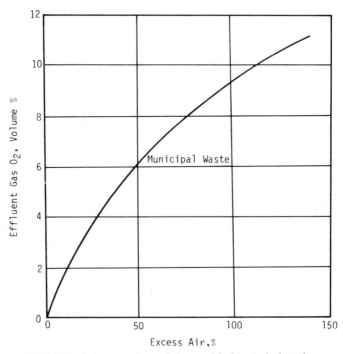

FIGURE 2.10 Oxygen vs excess air for municipal waste incineration.

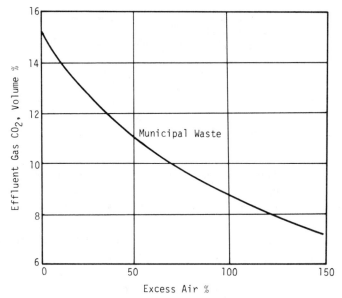

FIGURE 2.11 Carbon dioxide vs excess air for municipal waste incineration.

on the waste type). The high setpoint is the secondary chamber temperature required for complete combustion. At this point the auxiliary burner shuts off. Generally, the temperature setpoint on the secondary combustion air controller should be set approximately 100 degrees *above* the high setpoint for the auxiliary burner control to prevent "fighting" of the two control units. The control of primary chamber air is more difficult to control than secondary combustion air. Since the primary chamber operates to the left of the stoichiometric curve, the action of the controller is opposite to that of the secondary air controller. The underfire air controller would decrease air flow for temperature above the control temperature and increase air flow at temperature below it.

2.5 OPERATION BY VISUAL OBSERVATION

Incinerator Color

Fairly accurate determinations of air adjustments are possible by observing the incinerator color spectrum in both chambers of a controlled air incinerator. To use this technique, enclosed sight glasses must

46 CONTROLLED AIR INCINERATION

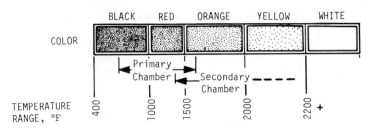

FIGURE 2.12 Combustion by color observation.

be mounted on each of the chambers. (If observation is made through an open door, localized air infiltration would negate the readings.)

It will be recalled that the volatiles in the waste are released from the waste in the primary chamber, and burned in the secondary. The carbon is burned in the primary. This activity is accompanied by a characteristic color pattern in the primary chamber. When the chamber is first loaded, the chamber appears black. As combustion proceeds, dull red flame begins to become visible, and then the color in the chamber gradually becomes dull red, and then dirty orange. This pattern (black, to dull red, to dirty orange) should repeat as each load of waste is added to the chamber.

If the color changes from dirty orange to brighter shades of orange or yellow, it is a sign that too much air is entering the primary.

Figure 2.12 shows the color patterns generally observed in the combustion chambers based on the procedures of McRee [2]. If contaminants are present, temperatures above 2000°F could cause rapid refractory deterioration. The color patterns described in Figure 2.12 apply to continuous, stabilized incinerator operation, and are not applicable to the heatup or burndown cycle of the operation.

Operational Problems

The three most common problems in controlled-air incinerator operation are excessive auxiliary fuel usage by the secondary burner, clinker problems in the primary chamber, and stack emissions.

2.6 AUXILIARY FUEL USAGE

In order to operate efficiently, a controlled air incinerator must maintain temperatures in each chamber within certain ranges specified by the design of the system and the waste being burned. If the heat content of the waste is not sufficient to maintain these temperatures, auxiliary fuel must be provided to make up the deficiency.

The most common cause of auxiliary fuel demand is inconsistent waste charging of the unit. Generally, if the heat content of the waste is greater than 5000 BTU per lb., it can be burned in a controlled air incinerator without auxiliary fuel demand (once thermal equilibrium is reached), *provided the waste is supplied to the unit at the design rate.* For example, a 1000 lb per hour unit must be fed 1000 lbs of waste per hour, at approximately 100 lbs every 6 or 7 minutes.

A second cause of auxiliary fuel demand is excessive air (either through leakage or improper control) in the primary. This results in combustion of some volatiles in the primary, reducing the amount of volatiles combustion in the secondary, lowering its temperature.

2.7 CLINKERS

Clinkers are formed in the primary chamber when the temperature remains at two high a level. These are masses of ash material which has "melted" and fused together, a condition which occurs at approximately 2200 °F and 2500 °F. Clinkers tend to adhere to the hearth and to interfere with ash removal.

The obvious solution to this problem is to keep the primary chamber temperature below 2500 °F, so that the ash will not soften and fuse. The difficulty here lies in the fact that the primary air is controlled by the temperature of the *chamber*, which is generally lower than localized temperatures at the primary air inlet ports. At these points, temperatures are often much higher than the "average" temperature of the chamber. *Often, bright yellow flame can be observed at these ports*, indicating high combustion temperatures.

The best method of handling this problem is to use small air inlet ports, thus minimizing the area of high temperatures, and at least reducing the size of clinkers which may form.

Another solution to the clinker problem is the use of the *combined water gas reaction* in which hot carbon combines with steam, producing carbon monoxide and hydrogen gas. These gases are burned in the secondary chamber, thus disposing of them. In the primary, the reaction is endothermic (takes on heat), resulting in minimization of localized hot spots at the air inlets.

2.8 REFRACTORY CONSIDERATIONS

2.8.1 General

Refractory requirements for incinerators vary widely. Refractory selection is governed by 5 factors discussed later—*chemical content, slag, heat, thermal shock and corrosion.* The refractory must be able to withstand occasional temperature excursions considerably higher than the design operating temperature.

The refractory material must be compatible with any slag likely to form on it. The nature of potential slag formation, including: acidity, major chemical components (iron, heavy metals, alkalies, etc), and the melting temperature of the slag must be known.

The presence of *chlorine or fluorine* will eliminate the use of castable refractories, since these elements will destroy the lime (CaO) binder.

Thermal shock is a rapid change in temperature caused by start-ups and shut-downs, increased feed rates, and runaway reactions.

Corrosion of the steel shell behind the refractory can also be a problem. Many corrosive elements can penetrate the refractory without damage and condense on the shell. However, three ways are discussed to help protect against corrosion of the steel shell.

2.8.2 Refractory Types

Refractories suitable for service in incinerators can be divided into three general types— brick, castables and plastics. Each is different and has its own unique advantages and disadvantages.

Brick refractories are power pressed or extruded into standard or special shapes. They are fired in the refractory manufacturer's plant to develop their strength. The incinerator can be placed in service as soon as the bricks are laid.

Castable refractories are mixed with water like concrete and poured (or cast) into the desired shape inside the incinerator. In larger incinerators, they can be applied by pneumatic (gunning) methods which are normally much faster than pouring. They develop their strength from a chemical reaction between a lime based binder and the water. Castable linings must be dried out before they are placed in service.

Plastic refractories come in a moist, moldable (plastic) form. Unlike castables, no water is added. They are installed in place inside the incinerator by ramming them with special air hammers. They develop their strength when they are fired prior to being placed in service.

2.8.3 Chemical, Thermal and Corrosion Concerns

Refractory requirements for incinerators vary as much as their designs. There is no such thing as a universal refractory recommendation for incinerators. What works well in one incinerator may be completely unsatisfactory in another.

Any refractory selection is limited by 5 factors—heat, slag, chlorine (or fluorine), thermal shock and corrosion. A refractory must be compatible with all of these limitations.

Any refractory installed in an incinerator should be able to withstand the maximum temperature inside the unit. Almost all incinerators are subject to run-away temperature excursions. This should be the limiting temperature, not the routine operating temperature. Also, remember that the refractoriness of any refractory is lowered by reducing atmospheres, slags, hydrogen and corrosive gases. Please read the article, "Refractories and Temperature," on pages 3 and 4 in *A.P. Green Castables—Instant Refractories*.

The refractory material should be selected so that it is compatible with any slag likely to form on it. It must be known whether the slag is acidic or basic, what the major components are (particularly iron, other heavy metals, alkalies, etc.), the melting temperature of the slag and any other information that can be obtained.

The presence of chlorine or fluorine will eliminate the use of castable refractories. These elements will attack the lime (CaO) binder and destroy it. This can cause refractory failures even at very low temperatures.

Thermal shock is a problem for most incinerators. This is a rapid change in temperature caused by start-ups and shutdowns, increased feed rates, runaway reactions and other causes. Some refractories can be severely damaged in short periods of time by rapid temperature changes while others are much more resistant to rapid temperature change.

Corrosion of the steel shell behind the refractory may have a major imput on refractory selection and design. Many corrosive elements can penetrate the refractory without damaging the refractory and condense on the shell causing serious corrosion. A good refractory design and material selection can reduce or eliminate this problem. There are three ways to protect against corrosion with refractory linings:

1. The refractory lining can be designed so that it will conduct enough heat to the shell to keep the shell temperature above the dew point of anything likely to condense on it. This is perhaps the best way to pro-

tect against corrosion; however, it results in considerable heat loss and can be a hazard to workers in the area.
2. An acid resistant lining can be placed between the shell and refractory. These acid resistant linings must be selected and installed with great care.
3. Very dense refractories with low porosity can sometimes reduce or eliminate penetration by corrosive components of the atmosphere inside the incinerator.

2.8.4 Refractory Selection

BRICK

Advantages
1. High density and low porosity—better resistance to slag penetration
2. Ceramic bonding—greater strength and resistance to destructive forces in their temperature range
3. Good hot strength in some types of brick

Disadvantages
1. More subject to damage by thermal shock than other types of refractory.
2. Slag attack and penetration at the joints.
3. Cost—generally brick are the most costly refractory because of the extra labor required to install them.

Recommendations
1. Lower temperatures and no slag problem—high duty or super duty.
2. Low temperature and slag—high fired super duty
3. Moderate temperatures and moderate slag—high alumina brick such as 70% alumina
4. High temperature and several slag—mullite bonded or alumina-chromia

These are just general recommendations. Specific recommendations may vary considerably from the general. It is advisable to consult the refractory manufacturer before making your decision.

CASTABLES

Advantages
1. The easiest and cheapest to install of all refractories
2. Good resistance to thermal shock
3. Some dense castables have good resistance to slag

Disadvantages
1. They lack the strength and slag resistance of brick and plastics
2. Castable linings must be prefired or heated very slowly on the initial start up
3. Binders can be destroyed by chlorine or fluorine

Recommendations
1. Low temperature and mild slag—super duty castables
2. Moderate temperature, moderate slag—high alumina castables
3. Moderate temperature, severe slag—chrome based castables
4. Higher temperatures—very high alumina castables

PLASTICS

Advantages
1. Excellent resistances to thermal shock (as a group, better than brick or castables)
2. Good slag resistance
3. Generally good compromise in properties between brick and castables.

Disadvantages
1. Must be fired in before placing the incinerator in service.

Recommendations
1. Lower temperature and moderate slag—super duty
2. Lower temperature, severe slag-phosphate bonded plastic
3. Higher temperature and moderate slag—high alumina materials
4. Higher temperature and severe slag—high alumina phosphate bonded or alumina-chromia phosphate bonded

This article is intended for general information only. Before selecting a refractory lining for your incinerator, please contact a refractory manufacturer to discuss your particular application.

2.8.5 Refractory Costs

It is very difficult to give a cost estimate for lining a typical incinerator. Costs vary according to the size of the unit, complexity and quality of refractory required. Listed below are some general price ranges for refractories commonly used in incinerators in 1983 dollars. Castables and plastics are stated in terms of dollars per board foot while brick prices are stated in terms of cost per 9 × 4 1/2 × 2 1/2 (the standard refractory brick).

1. Castables
   a. Super Duty $1.00–1.75
   b. High Alumina $2.50–3.20
   c. Very High Alumina $7.20–8.00
   d. Chrome-base $3.00–5.00
2. Plastics
   a. Super Duty $1.40–1.70
   b. Phosphate Bonded $3.10–4.90
   c. Very High Alumina, Phosphate Bonded $7.40–7.70
   d. Alumina-Chromia, Phosphate Bonded $11.50–13.50
3. Brick
   a. Super Duty $.80
   b. High Fired Super Duty $1.00
   c. 70% Alumina $1.70
   d. 90% Alumina, mullite bonded $6.10
   e. Alumina-Chromia $12.00+

These figures are for materials only. The cost of installing them varies considerably from unit to unit; however, a rough rule of thumb is that it takes approximately $1.00 of labor to install $1.00 of refractory. It is best to consult a firm specializing in refractory installation for any refractory construction cost.

> Ambient to 250°F at 100°F per hour
> Hold at 250 for 3 hours
> 250°F to 500°F at 100°F per hour
> Hold at 500°F for 3 hours
> Increase to operating temperature at 100°F per hour

Plastic refractory linings must be fired in to develop their strength. Plastics are normally fired in at 50°F per hour with holding periods at 250°F and 1000°F for 8 hours. If the unit is not to be put in operation immediately, a 24 hour holding period at operating temperature is desired before shutting down.

A typical fire in schedule for a plastic lining is as follows:

> Ambient to 250°F at 50°F per hour
> Hold at 250°F for 8 hours
> 250°F to 1000°F at 50°F per hour
> Hold at 1000°F for 8 hours
> Increase to operating temperatures at 50°F per hour

For a detailed fire-in schedule for a particular lining, it is advisable to consult a refractory manufacturer or a firm specializing in firing in refractory linings. Any time a lining is fired-in faster than recommended, you are running the risk of causing an explosion or otherwise seriously damaging the refractory lining.

2.8.6 Installation Time

This varies greatly with the size and complexity of the unit. The following is a rough relative comparison of installation times for different types of refractories (starting with the fastest).

1. Castables (gunned)—This is normally the fastest means of installing a refractory lining. The castables are blown through a hose under pressure and mixed with water at the nozzle and sprayed onto the surface. This can only work for vessels large enough for someone to get into them with his equipment and do the gunning. More material is required for this application because refractory particles bounce off the surface (rebound) and do not stick. Rebound losses require an additional 10 to 25% more material.
2. Castables (poured)—Pouring castables like concrete behind forms is normally the second fastest way to line an incinerator. If complex forms are required, ramming plastics may be faster.
3. Plastics—Ramming plastics with special air hammers is normally somewhat slower than pouring castables. Again, the size and complexity of the unit determine this.
4. Brick—Brick normally takes the longest time to install. Special cutting and fitting required by unusual surfaces can greatly extend this time. The advantage to the brick lining is that it can be placed in service as soon as it is installed, whereas castables and plastics must be fired-in.

The purpose of firing-in the castables is to drive out the water. There are many fire-in schedules for refractory castables. They range from 30 to 100°F per hour with designated holding periods of ½ to 1 hour per inch of thickness.

2.9 BURNERS

Incinerators require burners for start up and for supplimentary heat. There are three basic burner classifications for incinerators. These are:

(1) Gas burners.
(2) Oil Burners.
(3) Combination Gas-Oil Burners.

Each of these types can be divided into many sub-types which are classified by various methods such as mixing, draft arrangements, and atomizing systems.

2.9.1 Gas Burners

The types of gas burners can be divided into three basic categories:

(a) *Raw Gas Burners.* The raw gas burner is one in which the gas is introduced into the combustion air as a pure fuel at the point of ignition. This type is usually used where there is a high level of air supplied either by a natural or a forced draft system.
(b) *Nozzle Mix Burners.* The nozzle mix burner is essentially a raw gas burner except that mixing with air occurs at the end of a single injection nozzle within the burner. Air/fuel ratios are controlled as compared with a raw gas burner which doesn't attempt to control the air-fuel ratio. The nozzle mix burner usually employs mechanically linked valves or dampers to control this ratio, but may also use a "zero" regulator arrangement.
(c) *Premix Burners.* The premix burner takes several forms. The true premix burner generates a fuel-air mixture and pipes it to one or more burners. This system is usually used for multiple burners where a constant ratio is required on each of the burners. A major shortcoming of this type of burner is the potential for flame flashback from the point of ignition to the point where the mixture is first generated. It is not widely used due to this problem.

2.9.2 Oil Burners

There are many designs, modifications, and patents on oil burners. The simplest method of classifying them is on the basis of the method by which the fuel is made conbustible (i.e. changed from a liquid, which won't burn, into a vapor or unit which will burn). Before describing the various oil burner classifications, the basic components of an oil burner should be analyzed.

The first part is the liquid injector. The liquid must be transported into the furnace or combustion chamber, and it must be either finely divided or atomized or vaporized.

Atomization can be accomplished by forcing the liquid through a small opening (orifice) under air pressure; by spinning the oil in a cup or on a disc at a high rate so that it sprays off into the burner; or by combining the oil with a stream of compressed air.

The second part of the burner or combustor is a windbox. The windbox is the point at which air is introduced into the burner body and usually contains vaneous fan. Burners may be forced draft or natural draft depending upon the type of system into which they are firing. Forced draft burners are inherently more efficient and better combustors because they can achieve higher air velocities than natural draft systems. The natural draft units are dependent upon the height of the stack and therefore pressure drop is quite critical.

The third part of the burner is the combustion chamber. Most commercial burners utilize the downstream device into which they are firing as the combustion chamber and have only a refractory flame burner block or tile which supports the base of the flame.

Oil burners, therefore, as indicated above, can be divided into several categories either by the type of mixing or by the type of atomization. The first, the type of mixing, is either forced draft or natural draft. The atomization is either mechanical, low pressure air, high pressure air, steam, or sonic, and the mechanical atomization includes the rotary cup type of unit.

2.9.3 Combination Gas-Oil Burners

Combination burners permit the use of either gas or oil as a fuel. Use of combination burners makes it possible to take advantage of the economic breakpoint between gas and oil at various seasons of the year, and/or through price fluctuations of the two fuels. The combination burner, while more expensive initially, can offer significant savings during operation.

The combination burner is essentially an oil burner with the necessary adaptation to permit firing of natural gas or liquid petroleum. This usually involves a nozzle mix type of burner with arrangements as described in Section 4.3.1.

2.9.4 Ignition Systems

Most burners are spark ignited. This involves a spark plug similar in many ways to what might be found in the standard automobile except that the industrial spark plug has a higher voltage and a wider gap. The voltage normally employed for industrial burners is in the range of 6000 volts. An ignition transformer converts 120 volt input power to 10,000 volt secondary coil output.

2.9.5 Burner Controls and Safety Equipment

The safety equipment which is required on either an air, gas, or combination system is fairly closely specified by the major insuring agencies (such as Factory Mutual which gives a list of approved components and also specifies certain arrangements which are acceptable or which are unacceptable).

Generally, these require some or all of the following features:

(a) A requirement that the entire system, depending upon its downstream volume, be purged with air for a certain period before ignition of the gas pilot or main burner ignition. This is to insure that no combustible mixture exists within the incinerator.
(b) That there be a limited trial for ignition of the pilot after which it will be necessary to go back to a prepurge condition for the entire system.
(c) That the main burner ignite within a specified time period, usually in seconds, after the main fuel valve has been opened.
(d) That an approved type of flame failure device be utilized on the burner.

FLAME FAILURE DEVICES

Flame failure devices can be separated into three groups:

(a) Flame rectification devices (flame rod)
(b) Photo cells
(c) Photo-chemical active cells

Photo cells, flame rods, and flame rectification systems are not used widely.

The photo-chemical cell generates a very small voltage which can be amplified to hold open relays as long as the ultraviolet scanner senses a flame within the burner. Upon flame failure, the voltage ceases and the solenoid valve on the main gas line closes, shutting off the fuel flow to the burner.

AUXILIARY EQUIPMENT

In addition to the items described above in the control system for burners, such things as combustion air blowers for forced draft burners, normally called turbo blowers, are used to supply air to the system. Most forced fraft burners utilize 100% primary combustion air which must be supplied by fans or blowers.

Fuel oil pumps are normally of the gear pump type achieving pressures from 30–400 psig. The pressure is controlled by a relief valve on the discharge of the pump which recirculates some of the fuel back to a storage reservoir.

TABLE 2.1
Burner Btu rate for natural gas consumption.

Procedure:

Step 1— Read the gas meter at beginning and end of a one (1) minute interval.

Step 2— Multiply the meter reading difference by 60 to get cubic feet per hour (ft^3/hour)

Step 3— Multiply ft^3/hour by the pressure correction factor to get corrected ft^3/hour

Calculate Burner Btu's

Step 4— Multiply the natural gas consumption (ft^3/hour) by the Btu content (900-1100 Btu/ft^3) of the natural gas.

≈ ft^3/hour × Btu/ft^3 = Btu/hour

Example of gas consumption for natural gas burner:

Gas consumption after one (1) minute (ft^3/1 min.) = 15 cu. ft.

Ft^3/hour = 15 cu. ft./minute × 60 min./hour = 900 cu.ft./hour

Apply Correction Factor:

900 ft^3/hour × correction factor (10 lb gas pressure)

Ft^3/hour = ft^3/hour × 1.666 = 1,494

Burner Output:

1494 ft^3/hour × 1000 btu/ft^3 = 1,494,000 Btu/hour

Status of Burner

Manufacturer recommends primary burner be fired at 1,000,000 Btu/hour.

 Burner operating above the recommended firing rate.

2.9.6 Burner Rate Checks

Procedures for determining burner Btu rate are illustrated in the following tables with examples for calculating fuel consumption for gas and oil burners. Table 2.1 is for natural gas, Table 2.2 for LP gas and Table 2.3 for fuel oil.

2.10 SPARE PARTS

Minimum spares for a controlled air incinerator should include:

1. Hydraulic fluid
2. One set of hydraulic cylinder seals for each size supplied.

TABLE 2.2
Burner Btu rate for liquid petroleum (LP) gas consumption.

Procedure:

Step 1— Read the gas meter at beginning and end of a one (1) minute interval

Step 2— Multiply the meter reading difference by 60 to get cubic feet per hour (ft^3/hour)

Step 3— Multiply ft^3/hour by the pressure correction factor to get corrected ft^3/hour

Calculate Burner Btu's

Step 4— Multiply the LP gas consumption (ft^3/hour) by the Btu content of the LP gas (2550 Btu/ft^3)

≈ ft^3/hour × Btu/ft^3 = Btu/hour

Example of Gas Consumption for LP gas burner:

Gas consumption after one (1) minute (ft^3/1 min.) = 6 cu. ft.

Ft^3/hour = 6 cu. ft./minute × 60 min./hour = 360 ft^3/hour

Apply Correction Factor:

360 ft^3/hour × correction factor (10 lb gas pressure)

ft^3/hour = 360 ft^3/hour × 1.666 = 599.76

Burner Output:

599.8 ft^3/hour × 2550 Btu/ft^3 = 1,529,490 Btu/hour

Status of Burner:

Manufacturer recommends primary burner be fired at 1,500,000 Btu/hour.

 Burner is operating in the correct range.

---

3. Flame safety relay for burner(s).
4. Flame Sensor(s).
5. Limit switches or other position indicators.
6. Thermocouple of each type supplied.
7. Temperature controller of each type supplied.

2.11 DESIGN CONSIDERATIONS

Several modifications have been developed to improve the operation of controlled air incinerators. These include changes to the chamber, use of continuous feed systems, and steam injection for more complete carbon combustion. Table 2.4 summarizes some of these modifications.

A typical design, operating characteristics and utility requirements

TABLE 2.3
Burner Btu rate for fuel oil consumption.

Procedure:
  Step 1— Make initial reading of the fuel oil meter
  Step 2— Reread the meter after ten (10) minutes
  Step 3— Multiply the difference in the two readings by six (6) to determine gallon per hour (gph) oil consumption
  Calculate burner Btu's
  Step 4— Multiply the fuel oil consumption (gph) by the Btu content of the fuel oil (Btu/gal)

For No. 2 Fuel Oil
Gallons/hour × 142,000 Btu/gallon = Btu/hour
Example of Fuel Oil Burner Consumption Based on No. 2 Fuel Oil:
Initial Meter Reading—   5060 gallons
Reading After 10 Minutes—   5061.5 gallons
Oil Consumption:
5061.5 - 5060.0 = 1.5 gallons/10 minutes
Gallons/hour = 1.5 gallon/10 minutes × 6 = 9 gallons/hour
Burner Output:
9 gallons/hour × 142,000 Btu/gallons = 1,278,000 Btu/hour
Status of Burner
Manufacturer recommends primary burner be fired at 1,500,000 Btu/hour.

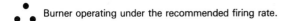 Burner operating under the recommended firing rate.

---

listing is given in Table 2.5 for Starved Air Incinerators. These values will vary with materials burned and method of operation. A survey by Frankel et al [3] shows that typical first chamber temperatures are 600–1600 °F and the second chambers are 1200–1800 °F at ± 0.2 inches water gauge pressure. The second chamber residence time is about 2 seconds.

2.12 PROBLEM WASTE OPERATION

The chemical and/or physical nature of some waste materials cause special problems with emission control, refractory selection, and primary chamber hearth design. Some of these wastes, and the associated problem areas are summarized in Table 2.6.

TABLE 2.4
Modifications to controlled air incinerator to provide better combustion and continuous operation.

Item	Purpose	Comments
Enlarged secondary chamber	needed for continuous operation of controlled air incinerator	some units have added checker work for turbulence
Continuous feed	Ram feed provides semi-automatic operation	Original units were batch feed—units were changed thru large door on end of unit
Steam injection	to aid in water shift reaction to complete combustion of char (carbon) in primary chamber	Also helps to prevent slag formation by completing reaction at a lower temperature
Ash removal	moving refractory panels push ash/non combustible to end of unit	ash falls into water seated chamber where it is quenched & removed with conveyor

TABLE 2.5
Typical design criteria and operating characteristics.

(A) *Design*

- Typical loading rates: 400-3,000 lbs/hr
- Heat release: 20,000 Btu/cu ft. (total unit)
- Excess air: 100%
- Normal operating temp.: 2,000-2,200°F
- Burn out: 99.94%
- Dwell time: 1.25-1.60 seconds
- Types of waste usually burned: Type 1,2, & 3
- Principal applications: Hospitals, schools, manufacturing plants, retail stores, apartment buildings

(B) *Operation* (rated per ton of refuse processed)

Operation Labor:

12.5 ton plant—	one (1) man-hour per ton
25 ton plant—	1/2 man-hour per ton
50 ton plant and larger—	1/4 man-hour per ton

Auxiliary fuel: 1.5-1.7 MMBTU per ton

Electric: 10 kw hours per ton

Water:
  For Incinerator   11 gallons per ton
  For Housekeeping, etc. 390 gallons per day

TABLE 2.6
Impact of problem wastes.

Type of Wastes	Effects Emissions	Effects Refractory Life	Effects the Hearth (i.e. slag, etc.)
1. Salts or oxides of titanium, cadmium, zinc, aluminum, tin, lead, antimony, bismuth, sodium, calcium, potassium, magnesium and barium	•		•
2. Phosphorous compounds	•		
3. Flame retardant materials	•		
4. Diatomaceous earth, vermiculite and similar materials	•		
5. Filter media	•		•
6. Metallized labels or wrappings	•		
7. Hazardous wastes	•	•	•
8. High silica content materials	•	•	•
9. Large quantities of any inorganic material	•	•	•
10. Paints, pigments, sludges, liquids	•	•	•
11. Some vegetable hulls	•	•	
12. Asphalt shingles	•	•	•
13. Dirt, sand, clays	•	•	•
14. Soaps and detergents	•	•	
15. Halogenated materials	•	•	
16. Sulfur bearing materials	•		
17. Materials with silicones	•		
18. Polymeric materials which melt before burning		•	•

2.13 MAINTENANCE

Maintenance consists of routine, pre-planned procedures as well as emergency procedures to predict and eliminate potential problems and to repair unforeseen failures. An EPA survey [4] shows that the principle routine maintenance for incinerators consists of small refractory repairs and replacement of thermocouples, switches, door seals and motors plus periodic:

removal of soot from boiler tubes
cleaning of induced draft fan blades
cleaning of internal boiler tubes
removal of slag from air injection ports.

Refractory in general was found to last about 3 to 8 years.

2.13.1 Typical Maintenance Procedures

To continue to obtain reliable service from the incinerator, it is necessary to perform periodic inspections and maintenance servicing. Table 2.7 summarizes some of the typical maintenance tasks.

More specific details on the typical servicing of the incinerator follows:

(1) CLEANING THE SPARK PLUGS

To clean the spark plugs on the burners, remove each spark plug assembly from the base of its respective burner. Clean the tip of the spark plug electrode and its grounding point using fine steel wool or fine sandpaper, removing all soot or carbon buildup. Reset the spark plug gap (the distance between the electrode and ground point) according to the specification provided with the equipment.

(2) CLEANING THE FLAME DETECTORS

Remove the device from the base of the burner by loosening the nut which attaches the flame detector to the burner plate nipple. Locate the lens inside at the end with the attachment nut. Clean this lens with a dry cotton cloth or tissue, removing any soot or carbon buildup. Replace the

TABLE 2.7
Maintenance schedule.

Component	Maintenance	Frequency
Ultraviolet Flame Detectors (Burners)	Clean lens	Monthly
Opacity Monitor	Clean light source and Receiver lens	Monthly
Fan	Grease bearings	Semi-annually
Burners	Clean Spark plugs	Semi-annually
Combustion Chamber	Check refractory for damage	Annually
Damper Motor	Lubricate	Annually
Fan	Clean Blades	Annually

detector on the burner pipe nipple by tightening the nut *"hand tight only."*

(3) FAN MOTOR LUBRICATION

The fan motor must be lubricated occasionally with an all-purpose grease.

(4) CLEANING THE FAN

The fan should be cleaned annually to insure proper air movement. To remove the fan for cleaning, disconnect the electrical power, remove the fan from its mounting, and visually inspect the fan blades. Dirt buildup on the blades should be scraped off. *Do not remove small balance weights located on some blades.*

(5) BURNER GAS ADJUSTMENTS

The burner gas settings for each burner should be checked and adjusted separately. This may be done by reading the fuel consumption with the burners running. Proper gas firing rates are very important to pollution free operation, extended life of the refractory and fuel conservation. To change the firing rate, adjust the manual gas valve on each burner. *This procedure should only be attempted by a skilled, trained technician or a gas company representative.*

(6) REFRACTORY INSPECTION

The refractory should be inspected and evaluated annually. *Be sure gas and electrical services are turned off. Two people should perform this service, with one remaining outside the incinerator to insure that the utilities are not accidentally turned on while inspection is in progress.* Inspect the refractory to make sure that the refractory has not fallen away, exposing any of the steel structure. The refractory is generally at least 4½" thick. General wear or spalling, of up to 2" depth can be tolerated, but should be noted for future repairs. Wear of spalling beyond 2" depth should be repaired as soon as possible. Consult factory for further instructions.

(7) STACK AND FAN MAINTENANCE

These procedures are detailed in Table 2.8.

2.13.2 Troubleshooting

The incinerator is a relatively simple device, and except for routine maintenance checks, the only areas where troubleshooting may be necessary are the blower and burner(s). Make sure the power supplies are turned off before performing any service. The following listing is a guide for troubleshooting blowers and burners:

64 CONTROLLED AIR INCINERATION

TABLE 2.8
Stack and fan maintenance.

Activity	Frequency	Remarks
Grease fan bearings	Semi-Annually	------
Check fan for excessive vibration	Daily	Vibration must be minimized immediately
Check fan belts for proper tension, wear	Daily	If belts loose, tighten immediately; replace if worn or cracked. Be certain to replace with proper size and type of belt—keep spares on hand
Thoroughly inspect stack for holes, cracks, leaks, etc.	Annually	Repair the stack as soon as practicable
Verify fan speed with a tachometer	Semi-Annually	If speed off, check for proper belts and sheaves, alignment and belt tension. Change of belts/sheaves may be required, or the motor may need replacement
Have the fan wheel spin balanced	Annually or when replacing	

TROUBLE SHOOTING THE BLOWER

If blower does not start:

(1) Check the power source, or supply Breaker.

(2) Re-set the motor starter by depressing the re-set bar.

TROUBLE SHOOTING THE BURNER

If burners will not fire:

(1) Check to be sure the safe run light is on.

(2) If safe run light is not on, check the loading door safety switch to be sure the door is in the correct position.

(3) Check the air safety switch to be sure the blower is providing sufficient air.

TROUBLE SHOOTING THE OIL BURNER

(1) *Flame away from the burner* . . . the most likely reason for the flame

being too far from the burner is that it is being pushed off by too much primary air. Check and clean the air nozzles and/or adjust the primary air according to the manufacturer's instructions.

(2) *A smoky flame* (unstable and flickering) means not enough air. The primary and secondary air controls should be lubricated, cleaned and reset according to the manufacturer's instruction. Fan blades may need to be cleaned and belts tightened.

Further burner trouble shooting procedures are summarized in Table 2.9.

2.14 ROUTINE OPERATION

Incinerator operational procedures must be established for every system. These must be tailored to each specific installation with its peculiar problems as every installation can be different. The procedures must be taught to the operators and a check method must be initiated to insure that the procedures are followed. Typical routine procedures presented by Cross and Hesketh [5] are listed below as examples.

Routine Start-up
1. Examine the unit carefully. Make sure it is clean, that the doors and interlocks operate properly, that both the primary and secondary

TABLE 2.9
Oil burner trouble shooting summary.

Flame Problem	Possible Cause	Areas to Check
Flame Away From Burner	Too Much Primary Air	Primary Air Shutter, Linkage, Fan
Smokey Flame	Not Enough Air	Too Much Oil Incorrect Cup Position
Flame Too Long	Too Much Oil Incorrect Cup Position	Oil Valves Burner Cup
Flame Too Wide	Too Little Primary Air; Incorrect Cup Position	Primary Air Shutter, Linkage Fan Burner Cup
Sparky Flame	Oversized Bits of Oil and Carbon	Cup—Clean, Possible Adjustment
Pulsating Flame	Oil amount incorrect Uneven Oil Flow Too Little Air	Oil Temperature Oil Pressure Air Supplies

burners ignite and burn cleanly, and that temperature and pressure instruments are set at the recommended set points.
2. Close all doors, start fans and purge with air for two minutes or longer if temperature is >300°F.
3. Ignite main burner and bring up temperature of primary chamber to recommended level over the recommended heating rate compatible with the refractory limitations.
4. Charge with 50 percent of normal waste charge.
5. Ignite secondary burner and adjust for clean stack.
6. Adjust air flow to recommended value and begin normal operation when specified temperatures are achieved.

Routine Operation
1. Charge at periodic intervals in accordance with operating instructions. Weight or volume measurements ±5 percent will ensure good incinerator operation provided the density of the charge is consistent.
2. Adjust air flow and temperature of afterburner chamber to achieve clean stack (visual). Excess air flow should be reduced to absolute minimum (or recommended value if measurement equipment is available) and afterburner should be a minimum fuel level to produce clean stack.
3. Continue charging at recommended rate, but decrease rate if an increase in unburned waste develops in the incinerator.

Routine Shut-down
1. Stop incinerator feed.
2. Determine that charge has been incinerated and only ash remains.
3. Shut down main ignition or support burner if still operative.
4. Shut down secondary burner and cool at the rate recommended for the refractory used.
5. Purge with air until temperature is reduced to 300°F.
6. Shut off blower and see that all interlocks are secured. If draft is induced leave the blower in operation.
7. Open charging and access doors and shut off induced draft blower.

In addition to routine procedures, "special" procedures should be developed for each facility.

One of the major areas of future operational concern, is the contingency and evacuation plan for the incinerator facility. Incinerators have both the potential of spills from charging containers and/or piping and the potential of detrimental by-pass emissions during malfunctions.

TABLE 2.10
Example outline table for isolation and evacuation distance.

Source Material	Initial Isolation In All Directions, ft		Initial Evacuation From Large Source		
	From Ground Level Source	From Elevated Source	All Directions, ft	Downwind Direction, Miles	
				Width	Length
Acrolein	550	N/A	1140	3.0	4.7
etc.	etc.	etc.	etc.	etc.	etc.
•	•	•	•	•	•
•	•	•	•	•	•
•	•	•	•	•	•

The contingency plan has to address both of these circumstances and the operator must be aware of the consequences of his actions as it affects the surrounding community.

DOT may publish a table of evacuation distances for different materials. The siter and operator of an incinerator facility must keep this in mind. Table 2.10 is an example table outline. The impact from the incinerator system must include incinerator failures, scrubber failures and leaks/spills as suggested in Table 2.11.

TABLE 2.11
Worst case estimates of emissions and resulting ambient air concentrations due to emergency problems.

Shutdown Due to:	Emissions, g/sec		Duration of Event, Minutes	Ambient Concentration of Pollutants, g/m³	
	Fugitive Particulates	Fugitive Organics		At Maximum GLC	At Property Line
1. Improper Incineration					
2. Scrubber Failure			(Fill in Depending on Facility)		
3. Unloading Operations Spill/Leak					
4. Tank Farm Spill/Leak					

REFERENCES

1 Cross, F. L. and Flower, F. B., "Controlled Air Incinerators," 3rd Annual Environmental Engineering and Science Conference, U. of Louisville (March 1973).
2 McRee, Robert E., "Operation and Maintenance of Controlled-Air Incinerator," 76th APCA Annual Meeting, Atlanta, paper #83-591 (June 1983).
3 Frankel, I., Sanders, N. and Vogel, G., "Survey of the Incinerator Manufacturing Industry," *CEP*, Vol. 79, No. 3, pp 44–55 (March 1983).
4 Higgins, G. M. and Kleinhenz, N. J., "Burning Trommeled Refuse in a Small Modular Incinerator: A Technical, Environmental, and Economic Evaluation," U.S. EPA, Contract No. 68-01-3889 (March 1982).
5 Cross, F. L. and Hesketh, H. E., "Operation and Maintenance for Industrial Incinerator Equipment," 76th Annual APCA Meeting, Atlanta, paper No. 83-59.4 (June 1983).

CHAPTER 3

Emissions, Controls, Energy Recovery *AND* Costs

3.1 INCINERATOR EMISSIONS

3.1.1 Stack Emissions

The big advantage of controlled-air incinerators is that proper operation produces relatively low particulate emissions. Typical rates are 0.08 to 0.10 grains per standard cubic foot, with rates as low as 0.04 gr/set in some cases.

Excessive emission rates are generally the result of one or more of the following causes:

1. High setpoint for secondary burner too low.
2. Excessive infiltration air.
3. Excessive negative draft in the primary chamber.
4. Excessive primary air.
5. Excessive secondary combustion air.
6. Waste characteristics prevent operation at design conditions.

Most controlled-air incinerators operate in the range of -0.10 to -0.25 inches water column in the primary chamber. This negative draft should be sufficient to prevent discharge of smoke and odors when the charging door is open. A greater negative draft in the primary chamber tends to lift particulates from the refuse bed into the discharge gas stream, resulting in visible emissions from the stack.

Air infiltration often occurs as leakage around the charging door. Infiltration air can be expected at approximately the rates shown in Table 3.1. In smaller units, this infiltration air can represent a significant portion of the total combustion air if the charging door is not closed properly.

Figure 3.1 shows typical stack conditions on a controlled-air incinerator. Causes and cures for these conditions are as follows:

TABLE 3.1
Charging door infiltration air rates.

Primary Chamber Pressure ("H$_2$O)	Infiltration Rates (ft^3/min per in^2 open area)
−0.1	8.0
−0.2	11.5
−0.3	14.0
−0.4	16.0

Case 1: Secondary chamber temperature too low. Increase high-setpoint for auxiliary burner.

Case 2: Dense black smoke caused by unburned volatiles. Steps:
1. Increase secondary air to maximum.
2. Reduce underfire air.
3. Increase high-setpoint on auxiliary burner (1800°F for cellulosic, 2000°F for synthetic wastes).
4. Check charging rate and waste content (volatile materials).

Case 3: Hydrogen Chloride is probably being generated by combustion of chlorinated wastes. Incinerator adjustment will not solve this problem.

Case 4: Steady white smoke caused by entrainment of micron-sized par-

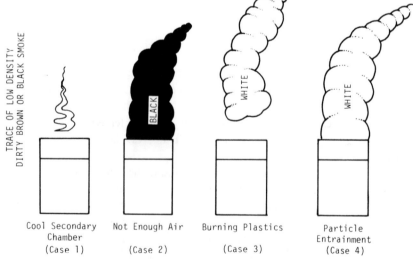

FIGURE 3.1 Typical stack conditions. (Courtesy R. E. McRee, Charlotte, NC)

ticles resulting from too much underfire air. Reduce underfire air flow. If condition persists, the cause is probably the characteristics of the waste.

Figure 3.2 compares typical air pollution standards and the emission levels of a controlled air incinerator.

3.1.2 Emission Factors

Recognizing the variations that can occur as discussed in Section 3.1.1, the U.S. EPA emission factors for incinerators [1] are listed in Table 3.2. The rating of "A" for this table means that the data are believed to be "excellent." Controlled emissions, other than those listed can be estimated by control efficiency values given in Section 3.1.4.

3.1.3 Particle Size Distribution

Distribution of particle size in incinerator emissions is reported by the McIlvaine Company [2] as shown in Figure 3.3. This indicates that the average mass mean diameter is large, being about 20 microns (μm) in size. The geometric standard deviation of these particles is also high at about 5.7.

3.1.4 Air Pollution Control

In addition to the improvements that can be achieved by proper operation as shown in Section 3.1.1 and as discussed in Chapter 2, emissions can be controlled by conventional devices. The U.S. EPA suggests [1] control device efficiencies as listed in Table 3.3. These values are only very approximate but serve to give a relative reference. Detailed descriptions of these devices, efficiencies and operating information can be obtained from Hesketh [3].

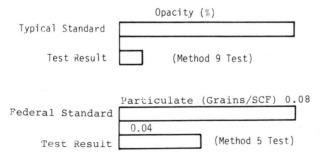

FIGURE 3.2 *Comparison of pollution standards and controlled-air incinerator emissions.*

TABLE 3.2
Emission factors for refuse incinerators without controls[a] — emission factor rating: A [1].

Incinerator Type	Particulates		Sulfur oxides[b]		Carbon monoxide		Hydrocarbons[c]		Nitrogen oxides[d]	
	lb/ton	kg/MT	lb/ton	kg/MT	lb/ton	kg/MT	lb/ton	kg/MT	lb/ton	kg/MT
Municipal										
Multiple chamber, uncontrolled	30	15	2.5	1.25	35	17.5	1.5	0.75	3	1.5
With settling chamber and water spray system[e]	14	7	2.5	1.25	35	17.5	1.5	0.75	3	1.5
Industrial/commercial										
Multiple chamber	7	3.5	2.5[f]	1.25	10	5	3	1.5	3	1.5
Single chamber	15	7.5	2.5[f]	1.25	20	10	15	7.5	2	1
Trench										
Wood	13	6.5	0.1[g]	0.05	NA[h]	NA	NA	NA	4	2
Rubber tires	138	69	NA	NA	NA	NA	NA	NA	NA	NA
Municipal refuse	37	18.5	2.5[h]	1.25	NA	NA	NA	NA	NA	NA
Controlled air	1.4	0.7	1.5	0.75	Neg	Neg	Neg	Neg	10	5
Flue-fed single chamber	30	15	0.5	0.25	20	10	15	7.5	3	1.5
Flue-fed (modified)[i]	6	3	0.5	0.25	10	5	3	1.5	10	5
Domestic single chamber										
Without primary burner	35	17.5	0.5	0.25	300	150	100	50	1	0.5
With primary burner	7	3.5	0.5	0.25	Neg	Neg	2	1	2	1
Pathological	8	4	Neg	Neg	Neg	Neg	Neg	Neg	3	1.5

[a]Average factors given based on EPA procedures for incinerator stack testing.
[b]Expressed as sulfur dioxide.
[c]Expressed as methane.
[d]Expressed as nitrogen dioxide.
[e]Most municipal incinerators are equipped with at least this much control.
[f]Based on municipal incinerator data.
[g]Based on data for wood combustion in conical burners.
[h]Not available.
[i]With afterburners and draft controls.

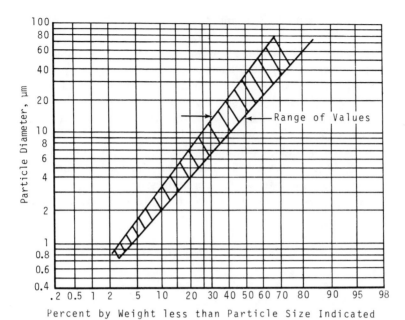

FIGURE 3.3 *Particulate size distribution of incinerator emissions prior to conditioning (courtesy of the McIlvaine Co. [2]).*

3.1.5 Emission Testing

Pertinent to the validity of emission factors and emissions is the accuracy and preciseness of the emission testing. This is not simple in many cases. For example, stack temperatures may be high or the gas may be saturated with water vapor. Testing of the latter can be achieved using

TABLE 3.3
Collection efficiencies for municipal incinerator particulate control systems.

System	Efficiency, %
Settling Chamber	0–30
Settling Chamber and Water Spray	30–60
Wetted Baffles	~60
Mechanical collector	30–80
Wet Scrubber	80–95
Electrostatic Precipitator	90–96
Fabric Filter	97–99

standard equipment and special precautions to keep the probe and filter dry.

High temperature sampling, however, requires special equipment. The currently accepted Method 5 sampling technique is good up to about 900 °F [4]. From 900 to 1200 °F a quartz lined probe could be used. Above 1200 °F a water cooled stainless probe is recommended to be used with quartz nozzles, an inconel sheathed thermocouple and an inconel pilot. The water cooled probe still must have a heated probe liner to prevent overcooling and condensing of the sampled gas vapors. Inconel metal is good to about 2500 °F but should not be used as a nozzle as it could spall and cause particulate error in the sample.

Any sample obtained should be corrected to 12% CO_2 (E_{12}) as follows:

$$E_{12} = \frac{(12) E}{\% \ CO_2}$$

where E = measured particulate emission rate
% CO_2 = CO_2 Orsat reading, in percent

3.1.6 Burning and Testing of Hazardous Wastes

EPA regulations require destruction removal efficiency (DRE) of at least 99.99% of the principal organic hazardous constituents (POHCs) in the waste [5]. The regulations for determining organic waste destruction are:

$$DRE = \frac{W_{in} - W_{out}}{W_{in}} (100)$$

where: W_{in} = mass flowrate of a POHC in the waste fuel
W_{out} = stack emissions rate of a POHC

In addition to POHCs, other requirements also exist for HCl and the criteria pollutants CO, NOx and HC. Products of incomplete combustion (PICs) are measures of total organics destruction but are not regulated directly.

Incineration is intended to destroy all POCHs while not producing harmful levels of PICs. PICs can be the more troublesome of the two in the atmosphere and the amounts of PICs can exceed the total of the original POCHs. PICs are defined as organic compounds not in the original waste feed that show up at $\geq$ 100 μg/g of feed during and after incineration operation.

An EPA Method 5 train modified according to U.S. EPA methods (40 FR, Part 60, Appendix A, 1980) for semivolatile organics (B.P. < 100°C) and particulate matter is shown in Figure 3.4. The EPA Method 6 train for HCl is shown in Figure 3.5.

3.1.7 Sampling and Analytical Procedures for Organic Emissions

Special sampling procedures are detailed for organic emissions in the summary by Polcyn and Hesketh [6]. This entire section consists of excerpts from this presentation.

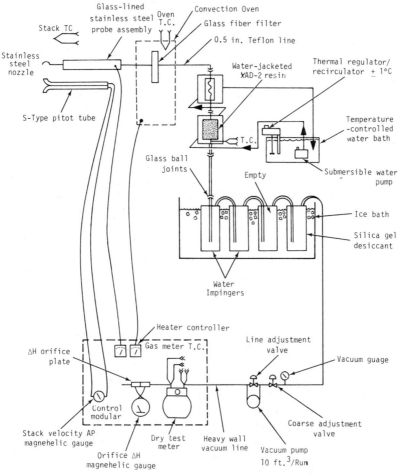

FIGURE 3.4 *EPA Method 5 train modified for semi volatile organics.*

76 CONTROLLED AIR INCINERATION

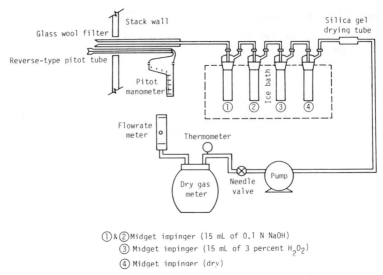

① & ② Midget impinger (15 mL of 0.1 N NaOH)
③ Midget impinger (15 mL of 3 percent H_2O_2)
④ Midget impinger (dry)

FIGURE 3.5 EPA Method 6 train hydrogen for chloride sampling.

SAMPLING METHODS

Table 3.4 presents a listing of selected sampling methods available to the user for various applications and compound classes. Although other methods exist that may have proven to be accurate and reliable, the ones presented were selected because they are the most commonly used, and/or are current methods that have been endorsed by the U.S. EPA and by the technical community at large.

The first three methods presented are essentially fixed-volume grab sampling methods. That is, the physical displacement of each dictates the maximum sample volume that can be collected. The first two methods (Syringe and Flow-through-bottle) are further limited in that the duration of sample collection usually lasts less than a few seconds. It is therefore considered an instantaneous grab sampling method.

The limited sample size (volume) that can be taken by these methods certainly limits the method detection limit, and the brief sampling period precludes obtaining a composite sample over any extended period of interest such as in sampling a cycling or non-steady state process emission source. They do, however, have the advantages of being low in cost and can provide very useful information in those instances when a short-term emission event needs to be defined.

The evacuated canister and Tedlar® bag methods are also volume limited, however, they do provide the ability to obtain a time-integrated sample. The canisters are of stainless steel construction and electropolished by the SUMMA process for inertness. Prior to sampling, they are cleaned by heating to 150 °C and by purging with ultra-high purity (UHP) air for at least 30 minutes. The canisters are then evacuated to approximately 75 mm Hg. Sample collection is flow regulated at a constant rate pre-selected to cover a desired sample duration. After collection of the sample the canisters are pressurized with UHP nitrogen to 10 to 12 psig. This provides an adequate positive pressure to allow for recovery of the sample from the canister. It also dilutes the oxygen and/or acid gas content of the sample for storage. This method is an improvement over the instantaneous grab approach in that time integrated samples can be collected, however, the sample size is still limited by the canister volume.

Sample collection by the Tedlar® bag method also provides the capability of collecting time integrated samples, however, it also provides the sampler with a much larger sample volume capability (6-, 12-, 20-liter capacity are typical) and the storage and recovery of the collected sample does not require pressurizing or dilution with UHP nitrogen.

Unlike the evacuated canister, the Tedlar® bag sample is collected by inserting the bag into an airtight rigid container and evacuating the container. The sample is drawn into the bag as the vacuum created in the container provides adequate suction to fill the bag. This procedure is referred to as Method 3 and is presented in detail in Appendix A of Title 40 Part 60 of the Code of Federal Regulations.

Although it provides a greater sample volume, there is no provision for concentration of the analytes of interest in order to achieve lower detection limits. Also, the bags are subject to absorptive losses of sample components and therefore should be analyzed onsite, as soon after collection as possible.

The sample collection procedure for EPA Method 25 is similar in principle to that for the evacuated canister. That is, the train consists of a polished stainless steel (S.S.) canister or tank that is evacuated to a very low absolute pressure (10 mm Hg). The sample is then drawn into the tank at a regulated flow rate. However, in addition to the tank, Method 25 employs a cold condensate trap in-series but prior to the S.S. tank to collect higher boiling point VOCs. This dual-staged approach provides a very useful split in the VOC range separating the very volatile compounds, which are mostly CO_2 and CH_4, from the less volatile compounds. This feature is an integral part of the sampling and analysis procedure required by Method 25 for determining Total Gaseous Non-Methane Organics (TGNMO) [6,18,21].

The sample recovery procedure involves a simultaneous purge and pressurization of the condensate trap and collection tank, respectively, followed by simultaneous oxidation of the trap contents to CO_2, catalytic reduction to CH_4 and transfer to and pressurization of an intermediate collection/storage tank. The analysis procedure simply involves transfer of triplicate aliquots (equal volumes) of the contents of each tank to the TGNMO analyzer for analysis by FID.

This method offers several advantages over the methods previously discussed, however, it still lacks the facility for concentration of the collected organics in order to provide useful, accurate results at sub-ppm levels. The lower detection level of Method 25 is typically 1 to 10 ppm depending largely on the interferences from H_2O and especially CO_2 [2].

TABLE 3.4
Sampling methods for toxic and hazardous organic materials [7].

Sampling Method	Description	Applicable Source Type	Applicable Compound Type	Applicable Analytical Method(s)	Sampling Method Limitations	References
Syringe	Instantaneous grab	—Non-combustion (storage tanks, spray booths, paint bake ovens, etc.) —low moisture content combustion emissions	Volatiles, C_1–C_{10}	GC-FID† or	Sample size & therefore detectable concentration are limited by container size. ≥1 ppm	7,15
Flow-through bottle	Instantaneous grab					
Evacuated Canister	Integrated grab		Volatiles, C_1–C_{10}	GC-MS** or		
Tedlar® bag (EPA Method 3)	Integrated grab	(example—boilers, dry control incinerators, etc.)	Volatiles, C_1–C_{10}	GC-PID††	Bag samples are subject to absorptive losses of sample components	
EPA Method 25	Two stage integrated grab train consisting of cold trap followed by evacuated S.S. tank	Non-combustion and low moisture content combustion emissions as above.	Volatiles & Semi-Volatiles, C_1–C_{16}	Oxidation/ Reduction to CH followed by GC/FID	Sample size is limited by tank volume. CO and HO can produce significant interferences. System is complex/ cumbersome	12,24,27
VOST*	Water-cooled sample gas, including condensate, is passed through dual in-series sorbent traps. Tenax G.C.® in first tube followed by Tenax G.C.® backed-up by charcoal in second tube.	Combustion emissions (boilers, hazardous waste incinerators, etc.)	Volatiles & Semi-Volatiles, C_1–C_{16}, Cl_1–Cl_{10}	GC-MS GC-ECD GC-PID	Sample size is limited to 20 liters per pair of sorbent tubes, Sorbent tubes are susceptable to contamination from organics in ambient air during installation and removal from train.	7,13,14,15

(continued)

TABLE 3.4
(continued)

Sampling Method	Description	Applicable Source Type	Applicable Compound Type	Applicable Analytical Method(s)	Sampling Method Limitations	References
Modified Method 5 (Fig. 3.4)	Water-cooled sample gas, with condensate is passed through single sorbent trap. Sorbent type dependent on compound(s) of interest.***	Combustion emissions as for VOST	Semi-Volatiles, PCBs, other halogenated organics C_7–C_{16}, Cl_1–Cl_{10}	GC-ECD, GC-HECD, GC-MS	Single trap system does not provide check for breakthrough. Flow rate limited to approximately 1 cfm.	7,10,15,17, 18,22,23,26
High Volume Modified Method 5	Sample gas is passed through condensers where moisture is removed before passing through two sorbent traps, primary followed by back-up. Flow rates of up to 5 cfm are achievable. Sorbent type dependent on compounds of interest.***	Combustion emissions	Semi-Volatiles, PCBs, other halogenated organics C_7–C_{16}, Cl_1–Cl_{10}	GC-ECD, GC-HECD, GC-MS	High flow rate results in high sampling train pressure drop requiring large pump capacity.	7,10,17,18 19,26

(continued)

TABLE 3.4
(continued)

Sampling Method	Description	Applicable Source Type	Applicable Compound Type	Applicable Analytical Method(s)	Sampling Method Limitations	References
SASS Train	Sample Gas passes through a cold trap followed by an XAD-2 sorbent trap. Train is all stainless steel construction.	Combustion emissions (boilers, hazardous waste incinerators)	Semi-Volatiles, and other non-halogenated organics C_7-C_{16}	GC-ECD, FC-HECD, GC-MS	System is complex, large and cumbersome. Recovery of organics from cold trap can be difficult. S.S. construction makes train components highly susceptible to corrosion from acid gases especially HCl	10,29

*VOST—Volatile Organic Sampling Train.
†GC-FID—Gas Chromatography with Flame Ionization Detector.
**GC-MS—Gas Chromatography-Mass Spectrometry.
††GC-PID—Gas Chromatography-Photoionization Detector.
***Sorbents include Florisil®, XAD2® resin, and Tenax-GC® among the most commonly used.

Method 25 has a few more limitations worth noting. These are, both the sampling and analytical apparatus is quite complex—this makes the method difficult to use and expensive; the sampling apparatus is also large and cumbersome; and the number of sample line connecting points, valves, etc. make meeting the leak check requirements difficult at best.

Some of the shortcomings of the manual Method 25 procedures can be overcome by the use of an automated analyzer version. The automated version follows the same oxidation/reduction steps in the sample analysis phase of the procedure as required by Method 25 but replaces the manual time-integrated grab sampling apparatus with a heat-traced Teflon sample line directly interfaced to the emission source and an automated TGNMO analyzer. The sample gas is drawn via the heat-traced line (maintained at a temperature above the dew point of the organics of interest) directly into the TGNMO analyzer which automatically separates the C_1 (methane) and nonmethane fractions by a series of chromatographic columns. Each fraction is analyzed separately using a timed temperature program using normal gas chromatographic procedures.

Care must be taken in selecting this automated version in that sample loss in the sample line and contamination of the chromatographic columns may result in erroneous results where very heavy organics must be measured (boiling point > 140°C). In such cases, the automated method may not be suitable.

The last four sampling methods commonly employ the principle of concentrating the organic constituents by using a sorbent trap or traps. This feature eliminates one of the short-falls of the previously discussed methods and that is, being able to achieve extremely low detection levels. This advantage rates the last four methods as the preferred (and possible the best) approach to determining the extremely low levels of toxic and hazardous organic emissions that must be measured accurately in the demonstration of the destruction and removal efficiencies (DREs) of various hazardous waste incinerators. Typically, hazardous waste incinerators must meet a DRE of 99.99 percent [7,15]. In order to demonstrate this level of destruction, the stack gas sampling method must be capable of collecting a significant quantity of sample gas and capturing and concentrating the principal organic hazardous constituents (POHCs) present in the gas stream. Although each of the last four sampling trains have their own limitations, they are well capable of achieving the necessary sample volume and concentration criteria to demonstrate compliance with DREs of 99.99 percent and greater. For example, the modified Method 5 sampling train has been successfully employed in demonstrating a DRE of 99.9999 percent [10,15,17,22] as required for destruction of PCBs in an Annex I incinerator, and the high volume modified Method 5 train was successfully used to demonstrate a DRE of 99.99999 percent for a high efficiency boiler burning a 5 percent blend of PCBs and mineral oil [7].

The only significant limitation of the standard flow rate (1 cfm) and high volume version (5 cfm) modified Method 5 (MM5) trains is the sample recovery efficiency achievable when using these methods. Although recovery efficiencies of 100 percent are achievable for certain compounds, the mean value covering the broad range of organics that these methods are used for is approximately 84 percent [1,4,9,16]. An additional limitation of the standard rate MM5 train is that, by design, it only employs a single trap. This does not provide a check for organics breakthrough, especially when large sample volumes (330 m^3) are collected. The high volume MM5 train employs a back-up trap, however, in so doing it imparts a very high headloss (20 to 25 in. Hg) that requires an extremely large capacity pump (17 cfm) to overcome and still provide the desired flow rate.

The Source Assessment Sampling System (SASS) train has been successfully used for Level I environmental sampling on numerous sources and is considered to be a useful method for the collection of semi-volatiles, PCBs and other lower volatility organics. However, the materials of construction employed in the SASS train (all stainless steel) make it very heavy and cumbersome and very susceptible to corrosive attack from the acid content of sample gases, especially HCl, from sources combusting high chlorine content wastes.

The VOST Method for collecting volatile organics is a relative newcomer to the list of tried and proven sampling methods. Although it was first introduced in 1980, [21] the first draft guidance document describing the VOST sampling and analytical procedures became available in April 1983 [13] and has only recently (February 1984) completed U.S. EPA's peer review process and should be available in final draft form in March or April 1984.

The VOST train has proven to be a reliable and accurate method for collection of the broad range of volatile and semi-volatile organic compounds [12,15]. By using a dual sorbent, dual in-series trap (tube) design, the VOST train virtually eliminates most of the limitations associated with the MM5 and SASS methods. However, it also has several inherent limitations that must be noted. These are, it is limited to a flow rate of 1.0 liter per minute (LPM) and a total sample volume of 20 liters per tube pair, therefore requiring frequent change-over of the tube pairs for tests that exceed 20 minutes; the frequent change-over of tube pairs makes the samples more susceptible to ambient contamination and to loss from breakage due to the amount of handling required. Tube change-over can be reduced by lowering the sampling rate to 0.25 LPM, however, this reduces the total quantity of sample by nearly 4-fold for the same sampling duration at 1 LPM.

In addition to the considerations discussed above, each of the last four methods listed in Table 1 have another common concern. That is, selection of the best sorbent material for a specific source or condition. Although the SASS train specifies use of XAD-2® sorbent and the VOST train uses Tenax-GC® with a charcoal back-up, other sorbents have been used in the MM5 and high volume MM5 with good success. Florisil®, Spherocarb and other polymeric sorbents as well as other petroleum-based sorbents have also been used successfully in both source and ambient applications [7,15]. Each sorbent has its limitations in what class of organics it is best suited for efficiently collecting and how efficiently the collected analytes can be recovered. Although no single sorbent can be considered "ideal" for all applications, the VOST train sorbents appear to have the best combined efficiencies for most applications, especially since both can be subjected to either thermal desorption or liquid extraction.

ANALYTICAL METHODS

The analytical methods listed in Table 3.5 are presented on the basis of their present and continued demonstration of successfully being employed by the analytical community, especially by the U.S. Environmental Protection Agency (EPA), for the analysis of organics in liquid and solid media as well as in air [7,15,25,30].

Because of the high degree of complexity in the field of organic chemistry, the only suitable approach for accurate, specific, and confident identification and quantitation of organic compounds is by either gas or liquid chromatography. However, the need for compound selectivity and specificity has led to the development of numerous detection methods. Each detector is based on a different operating principle and therefore each has its own advantages and disadvantages and further each has a unique response to different compound classes. Some detectors can be "fine-tuned" to a particular class or even sub-class of organics, such as the Hall Electrolytic Conductivity Detector (HECD), which can be operated in either the halogen or sulfur mode [15]. Unfortunately, most detectors are not "tunable" and depend on the chemical or electrical characteristics of the organic compounds for detection. Because of the inherent limitations found in every detection method, each method should be thoroughly understood before attempting its use in any given application.

Since the response of each detection principle varies by compound class, Table 3.5 presents the analytical methods listing their compound compatability or their ability to detect certain classes of organic compounds and gives the median instrumental detection level for the range of compounds given. It should be noted that the table is intended only as a guide.

The use of gas chromatography with a flame ionization detector (GC-FID) has been widely used for analysis of various organic compounds. The FID is based on the principle of "carbon counting," however its response to different compounds varies and therefore its use as a total hydrocarbon analyzer is questionable unless the compounds of interest are known and no more than three compounds are present in the sample [27]. When used to analyze specific compounds, if it is calibrated with a standard containing the specific compound of interest, it is a very accurate tool. EPA Method 25 employs an FID to determine total non-methane organics by converting the VOCs to methane and then analyzing the entire sample as methane. This eliminates the problem of variable response.

Unlike the other detection methods listed in Table 3.5, gas chromatography with mass spectrometry (GC-MS) is a highly sophisticated instrumental method whose operation is *independent* of the chemical or electrical characteristics of the compounds. It identifies and quantifies individual compounds by their mass spectra. That is, the gas chromatograph separates the compounds and the MS instru-

TABLE 3.5
Analytical methods for toxic and hazardous organic materials.

Analytical Method	Compound Applicability	IDL* (pg/m³)	Sample Preparation	Method Notes	References
Gas Chromatography (GC) —Flame Ionization Detection (FID)	Non-halogenated VOCs† Polynuclear Aromatic Hydrocarbons (PAH)	5-10	Direct injection, liquid-liquid extraction	Response varies with different compounds—not suitable for mixtures of numerous (>3) compounds	7,13,15,19, 24,25,27,30
	Acrolein, Acrylonitrile, Acetonitrile	25-100			
—Photo-Ionization Detection (PID)	Aromatic VOCs	0.1-1	Direct Injection	Excellent field screening method but at higher detection level	7,15,25,30
—Hall Electrolytic Conductivity Detection (HECD)	Halogenated VOCs	1-10	Soxhlet extraction, Purge and trap	Very halogen specific. It is capable of achieving very low-detection levels even when mixtures of numerous chlorinated compounds are present.	7,10,15,17, 25,30
—Electron Capture Detection (ECD)	Chlorinated hydrocarbons, Polychlorinated Biphenyls (PCBs), Organochlorine pesticides, cycloketones, Phthalate esters, Nitro-aromatics	1-1000	Soxhlet extraction, Purge and trap	Also highly halogen specific.	7,10,15,17, 25,30

(continued)

TABLE 3.5
(continued)

Analytical Method	Compound Applicability	IDL* (pg/m3)	Sample Preparation	Method Notes	References
Gas Chromatography-Mass Spectrometry (GC/MS)	VOCs, Semi-VOCs, PCBs, Halogens, PCDDs, PCDFs, etc.	100-1000	Soxhlet extraction, Purge and trap	Ideal for identifying and quantifying individual compounds in a mixture of numerous compounds.	7,10,13,14, 15,17,18,22 25,30
High Performance Liquid Chromatography (HPLC)**	PAH	0.1-1	Liquid-liquid extraction, soxhlet extraction	Highly specific for certain polynuclear aromatic hydrocarbons	15,25,30
Atmospheric Pressure Chemical Ionization Mass Spectrometry (APCI-MS)	VOCs, Semi-VOCs, PCBs, Halogens, PCDDs, PCDFs, etc.	100	Direct injection	Can be mounted in mobile laboratory for onsite analysis. Mobile capability has been demonstrated.	16,20

* IDL—Instrument Detection Limit; values given are ranges based on the median value for the range of applicable compounds listed.
† VOCs—Volatile Organic Compounds.
** With flourescence detector.

ment identifies each compound by its mass characteristics. The identification is made automatically by a computer scan of a complete library of the mass spectral data of thousands of compounds. This rapid and automated ability to accurately identify compounds by mass character rather than by chemical composition, makes the GC-MS instrument a very powerful tool for use in environmental studies. This is especially true with combustion air samples where various organic homologues or artifacts may be formed that are chemically difficult to identify and whose response by any other detection method could be misinterpreted leading to false positives or no detection at all.

Although the normal detection level that can be achieved by a GC-MS instrument is not as low as by other GC detection methods, it can be improved significantly by operation in the selective ion monitoring (SIM) mode. In this mode it examines the ratio of the major ion or ions of a suspected compound to that of a known standard in order to confirm its presence. The SIM mode of a GC-MS instrument allows detection levels to the picogram (10^{-12} gm) level.

High Performance Liquid Chromatography (HPLC) was listed in the table primarily because it has been demonstrated to be a very accurate and reliable method for certain applications of organic compound analysis [26,30]. However, the majority of experience with this method has been with samples collected from liquid and solid medial. Therefore, its application to samples collected in air should be carefully reviewed in light of the various alternative methods that have a much larger experience base.

As mentioned early in this section, the analytical community continues to constantly develop new methods of detection or to improve on already well established analytical methods for identification and quantitation of organic compounds. Such an improvement is listed in Table 3.5 as Atmospheric Pressure Chemical Ionization Mass Spectrometry (APCI-MS). This approach to the use of the mass spectrometric detection method ionizes the sample molecules by introducing them to the ion source in a stream of carrier gas where they react with ions generated from an atmospheric pressure corona discharge. The ions are sampled through a small orifice into a cryogenically pumped vacuum chamber and analyzed with a quadruple mass spectrometer.

The primary advantage of this approach is to permit real time measurement of nearly any class of organic compounds in air. Unlike traditional analysis by GC-MS, the sample gas is directly introduced into the APCI-MS system and is immediately analyzed onsite. This development is of significant importance to the analysis of orgnic compounds in ambient air, especially in applications such as characterization of ambient air around hazardous waste sites or at spill sites. It has, however, also been demonstrated as a practical approach to measuring trace organic compounds in combustion emissions [16,20]. A mobile version of the instrument, called by its manufacturer as the TAGA system (Trace Atmospheric Gas Analyzer), has been developed and tested in numerous applications over a two year period [10,14].

Based on the comments made in this section, selection of the appropriate detection method is of key importance to the proper analysis of any collected samples. This selection process should include the following considerations:

— Determine which compounds are of greatest interest.
— Determine what level of detection is needed or desired.
— Determine the minimum sample size required to achieve the desired detection level.
— Determine what interfering compounds may be present in the source emissions.
— Determine what sampling methods are best suited to the source being studied that are also compatible with the preferred analytical method(s).

QUALITY CONTROL

It is especially important that appropriate quality control procedures be employed throughout both the sampling and analytical phases of any project or study, because every study or test of any environment represents a unique situation. Quality Control (QC) must be present in all phases of any given program. That is, the staging or pre-field phase, the field sampling phase, the transportation phase (to and from the field study site), in the analytical phase, and finally in the calculation and reporting of results.

Pre-field Quality Control—Because of the potential for contamination of sample collection media, including glassware, sorbents and recovery apparatus, special care must be taken to properly clean and protect such items. Rigorous cleaning and conditioning procedures are documented in the literature [7,15,25,28,30] and should be applied to the extent that will ensure a level of quality needed to achieve the objectives of the study. Washing, rinsing, oven baking, sealing with clean and intert materials, and proper packaging for storage and shipment should all be considered.

Field Effort Quality Control—The potential for contamination or sample loss in field situations is very high. Therefore, taking adequate precautions to minimize their effects is a must to the successful completion of any sampling effort. Collection of duplicate samples, employing field blanks, control samples, and proper labeling, handling and storage of samples should all be included in the field QC program.

Quality Control in Transportation—In order to assure the safe transfer of sample containers both to the test site and in return to the laboratory, it is essential to always maintain complete chain of custody for all sample containers before and after collection. A custody transfer form should be used that clearly documents the custody, location, method of transfer, time and date of transfer and a description of the shipment (number of containers, size, type, if temperature controlled, etc.). In order to assure that samples are not confused with other samples or blanks, a well designed labeling system should be employed. Also, a packing slip should be included with each shipping container to further document sample custody and assure that actual contents match intended shipment. Another QC measure to consider is the use of trip blanks in the shipment process. This provides an added check for potential contamination that may occur from the handling, packaging and shipment of the samples.

Laboratory Quality Control—Laboratory quality control should include three basic areas. The first is in sample handling and storage, the second in sample preparation (extraction, clean-up, etc.), and the third in sample analysis. Just as for field QC, laboratory or method blanks should be used to evaluate contamination and artifacts that may be derived from glassware, reagents and sample handling. To assure proper QC in sample preparation, again method blanks should be used. Also, surrogate spiking compounds should be used to evaluate analyte recovery efficiencies. For analytical QC, surrogate spiking compounds will provide a good check on detection capability of the instrumental method being used. Other QC checks include interlab verification or splitting of samples (where sample size permits), the use of internal standards, and calibration of the analytical instrument(s) both before and after analysis of the sample series.

PROTOCOL FOR HAZARDOUS ORGANIC EMISSIONS TESTING

Each emission source type is unique and requires source specific elements in the development of a successful test program protocol. However, every test protocol contains basic elements that are common to all sources. These are definitions or descriptions of:

1. the primary program objective and any secondary objectives;
2. the test conditions;
3. the test schedule (number and duration of test runs, and the date and start time of each);
4. all sampling locations;
5. the number, size, and frequency of all samples to be collected at each location;
6. all sampling equipment/methods/procedures emphasizing application to the specific source type including any special modifications;
7. the proposed reporting of test results including methods of calculation and any assumptions;
8. the proposed quality control program including test sample custody transfer procedures and documentation.

3.2 ENERGY RECOVERY

Recovery of incineration energy and recovery of materials can provide large financial rewards and should be considered for nearly every type of incineration process. It is relatively easy to construct energy recovery facilities as noted by the 21 month completion schedule for the Portsmouth, NH system [31]. Hazardous waste incineration is no exception

FIGURE 3.6 *Three 2000 #/hr each municipal solid waste (MSW-Type 2) general incinerators with 3 cubic yard hydraulic loaders, heat recovery boilers (steam) and automatic ash removal; comptro type A-48 (courtesy John Zinc Co., Comptro Div.).*

[32]. Figure 3.6 shows a 3 incinerator system with heat recovery for municipal waste energy recovery.

3.2.1 Heat Recovery

The maximum heat recovery efficiency for Type O waste, 8500 Btu/lb, is about 67.9% based on the assumptions shown in Figure 3.7. In actual incineration the ash leaving the incinerator contains combustible material. Typically these ashes may have 40% ash and 60% combustible material.

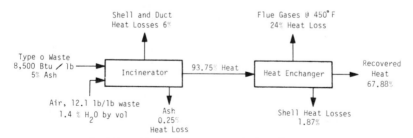

FIGURE 3.7 *Idealized heat recovery incineration system.*

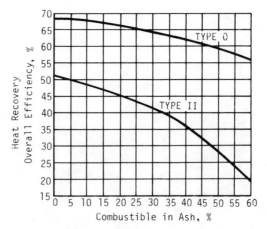

FIGURE 3.8 Heat recovery in controlled air incinerator as a function of fuel type and combustibles in the ash discharge.

For controlled air incinerators, this reduces actual heat recovery to 55% which currently exists. This is shown in Figure 3.7 along with the results of repeating the calculations for various percentages of combustibles in the ash. Shown in Figure 3.8 is the calculation results from burning lower heating value fuels such as Type II waste. Changing the combustion conditions will also change heat recovery. This is shown for excess air in Figure 3.9.

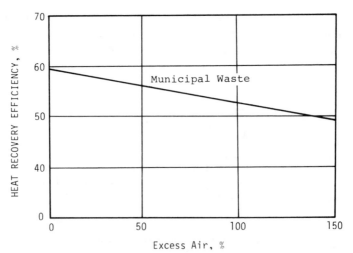

FIGURE 3.9 Heat recovery efficiency versus excess air.

Clearly, the controlled air incinerator is a tremendous heat source. The utilization of this heat for other purposes deserves high priority in today's market. Two common methods of utilizing this energy are steam production in a waste heat boiler and heating of air (for use as combustion air or for space heating), through a heat exchanger.

Most difficulties in waste heat boilers are found in the refractory and boiler tubes. The refractory tends to deteriorate due to temperature excursions, and rapid start-up/shut-down. Operating the incinerator below the dewpoint, improper purging, and failure to preheat the unit are a common cause of corrosion. Soot build-up, scaling, and slagging cause plugging of boiler tubes.

Indications of combustion efficiency and of heat recovery can be obtained using a few simple tools and techniques. Fuel mass feed rate and heating value are necessary starting points. In addition, the instruments listed in Table 3.6 are basic necessities to establish information as noted.

Static pressure measurements indicate if there is enough draft in the system. It also can help you evaluate the air handling system to determine such information as:

(a) Resistance in an undersized duct.
(b) Resistance in sharp turning elbows.
(c) Resistance across dry and wet collectors
(d) Static pressure of fans.
(e) Blockage of any sort in ducts or collectors.

TABLE 3.6
Basic trouble shooting instruments.

Measuring Device	What It Does	Purpose
Amprobe	Measures motor current/amperage	Indicates if fan is handling the proper amount of gases
Manometer	Reads static pressure	Determines if there is sufficient negative pressure in incinerator
Stem Thermometer	Measures gas temperature	Relates to combustion efficiency
Orsat or Fyrite Indicators	Checks gas composition (oxygen and carbon dioxide)	Relates to: —combustion efficiency —excess air

TABLE 3.7
Drum economics of open head steel drums.

	U.S. Department of Transportation	
Type of Expense	Rule 40, Misc. Non-Regulated	Rule 17H Regulated*
Cost of Purchasing New Drums	$12.50–$18.00	$24.00–$25.00
Cost of Purchasing Reconditioned Drums	$8.00–$12.50	$15.50–$17.00
Service Charge (Cleaning Cost Only)	$7.00–$10.00	$7.00–$10.00

*More stringent drum construction specifications.

(f) Incinerator draft problems.
(g) Fan Horsepower.
(h) Stack exhaust temperatures.

Additional information relative to fans and dampers for incinerators is given in the Appendix.

3.2.2 Material Recovery

Incinerators are used to recover materials by burning off unwanted contamination or coatings and preparing the basic material suitable for recycle and reuse. An example is the recycling of drums by use of incineration. Approximately 2/3 of the 75,000 drums used annually in the U.S. are reconditioned drums [33]. There are about 250–300 reconditioners at present. Table 3.7 is an example of drum reconditioning economics. It also implies that energy is saved because of this re-use aspect. Another equally important consideration is that recycling results in a reduced detrimental environmental impact. Purchasing of drums containing residue is no longer allowable under RCRA. However, drums with 1 inch or less material are presently exempt from regulation (the "1 inch rule") so they are not manifested at the recovery site.

RECLAMATION FURNACES

A typical drum reclamation furnace handles 250 drums per hour and could be classified as a hazardous waste incinerator as it *may* burn upwards of 600 pounds/hour of hazardous waste. Because of being *called* a reclamation furnace and not an incinerator, a RCRA B incinerator permit is not presently required. The result of this permitting shortcoming is that drum reclamation furnaces do not have to meet the

TABLE 3.8
Environmental concerns.

Process Step	Area of Concern	Potential Impacts
Delivery Area	Roadways	fugitive emissions from unpaved roads
	Drums being stored in trailers	fires and explosions
	Spills	runoff, atmospheric emissions, fire hazard
Drum Storage	Housekeeping	seepage runoff, odors, fumes, spontaneous combustion
	Containment	runoff/leachate into ground water source of fugitive VOC emissions, fumes
Drum Emptying Prior to Incinerator	Containment	runoff/leachate into ground water source of fugitive VOC emissions, fumes
	Sludge Handling	runoff/leachate into ground water source of fugitive VOC emissions, fumes
Drum Reclamation Furnace (Incinerator)	Operation (i.e., start-up, shut down & malfunction	visible emissions excess toxic emissions excess particulate emissions
Reconditioning Plant	Operation & Maintenance of baghouse serving shot blasting operation	visible emissions excessive particulate emissions
	Drying ovens O & M	excessive emissions of VOC's
	Spray Booths O & M	excessive emissions of particulates
	Process Water	spills going into storm drains effluent discharge not being properly treated sludge disposal from clarifiers

stringent destruction removal efficiencies (99.99% Destruction Removal Efficiency) of hazardous waste incinerators. Table 3.8 summarizes some of the concerns of reclamation facilities.

The drum reclamation furnace is a *tunnel* incinerator constructed with a secondary combustion chamber and, in most instances, a waste heat

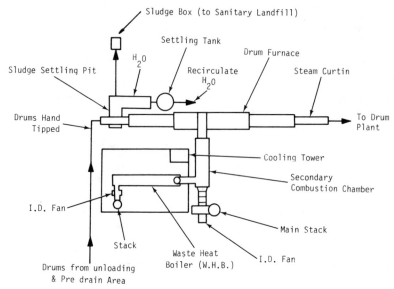

FIGURE 3.10 Typical drum incinerator flow diagram.

boiler. This is shown in Figure 3.10. There is one series of burners along the tunnel (primary combustion chamber) and a second series of burners in the secondary combustion chamber. All of the combustible material is supposed to be burned out of the drums as they pass through the incinerator. The off-gases (volatile matter) from the tunnel pass through the secondary combustion chamber where the combustion is completed. The hot gases then split with a portion (usually 20–30%) going through the waste heat boiler to generate steam and then out a stack, with the major gas flow (70–80%) going out the main stack.

Good burn out for the drums is achieved when the drums are carefully drained, the drum feed is not exceeded and the temperatures in the primary and secondary combustion chambers are properly maintained. All organic material is removed from the drums and there should be no visible or other unusual emissions. Table 3.9 lists typical reclamation furnace operating conditions.

In addition to the drum feed rate such as shown in Table 3.9, the type of material residue left in the drums is important. Table 3.10 lists some materials that might be encountered in drums being reconditioned in a typical drum reclamation facility. Each plant may handle different materials and percentages of drums in these categories.

TABLE 3.9
Typical process conditions for drum reclamation furnaces.

Parameter	Condition
Primary Combustion Chamber Temperature	1400°F
Secondary Combustion Chamber Temperature	1700°F
Retention Time	0.5 second
Chain Speed	10 ft/min.
Throughput	250 drums/hr.
Operating Schedule	10 hours/day 5 days/week 50 weeks/year

3.2.3 Design Effects

All of the energy and material recovery conditions are dependent upon good system design. A thorough check list for this prepared by Basic [34] is presented as Table 3.10.

3.3 COSTS

3.3.1 Incineration Systems

Typical capital and operating costs of incineration systems are shown in Figure 3.11 in 1984 dollars. These data are provided by Simmonds Manufacturing Corporation. Operating costs assume system is operating at designed rate and uses continuous heat recovery. Average values are used and fluctuations due to fuel heating value and densities and steam requirements are disregarded. Ash disposal could be calculated at

TABLE 3.10
Typical drum mix during testing.

Constituent	Mix of Drums (%)
Paint	30
Grease	5
Adhesive	10
Ink/Resin	5
Food/Cutter	30
Undercoat/Sealer	20
Total	100

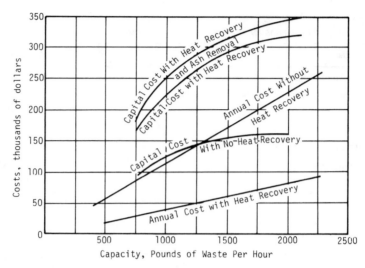

FIGURE 3.11 Capital and operating costs of incinerators with and without waste heat recovery in 1984 dollars (data courtesy of Simonds Manufacturing Corp.).

$35/ton and a mass reduction of 90% could be assumed from feed to ash, however, no waste disposal charges are included in these costs. Labor costs are based on one 8 hour shift per day, 52 weeks per year, 40 hours per week per operation and 1 operator per 8 hour day. Maintenance is estimated at 5 years for major repairs plus routine daily maintenance. Costs are calculated at $6.00/hour for labor, $0.05/KWH for electricity and $0.50/THERM for natural gas. Type 0 waste at 8500 Btu/lb is the fuel. Steam is produced from the waste heat at 120 psig from 212°F feedwater with gas cooling from 1800° to 450°F at an efficiency of about 55 to 58%. Value of the steam is $7.81/thousand pounds of steam. Figure 3.11 shows how significant the value of recovered energy can be compared with the other systems.

Another cost example [35] for hospital energy costs suggests that in 1972 hospitals consume 10% of all the commercial energy required in the U.S. Also, for typical hospital size incinerator operation the cost is $20/hour. Average hospitals generate daily up to 15 pounds per patient bed of high Btu waste and the incinerator units handle up to 19,000 pounds/day of this material. It is reported that waste recovery could provide up to 20% of the total hospital energy needs.

3.3.2 Air Pollution Control Systems

Costs of air pollution control (APC) systems can vary widely and specific estimates should be made for the particular types of system desired. In addition, many new types of systems (e.g., the low energy, high efficiency Catenary Grid Wet Scrubber) are now available at low capital and operating costs. For relative comparison, the McIlvaine Company [2] report costs of APC systems. These costs are given in Table 3.11 based on a 200 ton per day unit with 200% excess air at 600 °F (equivalent to 100,000 acfm). The costs are adjusted from 1969 to 1984 dollar values as dollars per 24 hour day with 2 shift per day operation.

TABLE 3.11
Checklist for selecting industrial waste-to-energy combustion systems (courtesy Basic Environmental Engineering, Inc.).

System Components	Potential Problems; Operating Experiences
1. *Preparation Equipment* a. Shredders, hogs, pulverizers	a. Best results with pure wood and paper waste products, but abrasive properties in even those waste requires maintenance.
	a. Use with wastes other than wood and paper increases problems with maintenance and downtime. Larger units may not be economically justifiable. Solvent containers present explosion hazards which must be taken into account in designing for safety.
b. Conveyors, Storage compactors, storage silos	b. Another mechanical component requiring attention. Care must be made to employ non-bridging designs of waste storage devices.
2. *Feeder Devices* a. Burners for pulverized wastes	a. Care must be taken to avoid explosions of powdered materials at pulverizer, storage and at burning chamber. Gun type burners are limited to burning wastes with less than 15% moisture.
b. Screw conveyors & underfeed stokers	b. Fairly reliable for granular materials. Primary caution is to prevent flame from propagating backwards.
c. Bulk waste ram type feeders, using hydraulic components	c. Must have trained persons who understand hydraulics. Minor improper adjustments to relief valves cause broken components and system shutdown. Broken hydraulic hoses and components can spread oil to burning chamber and create damaging fires.
d. Electromechanical bulk ram feeders	d. More than 12 years experience shows little to no maintenance required when these are equipped with fluid slip clutches.

(continued)

TABLE 3.11
(continued)

System Components	Potential Problems; Operating Experiences
e. Compactor-type feeders	e. Same problems possible as 2c above. When waste is chute fed, fire hazard results from flames propagating backwards.
3. *Burning Surfaces and/or Stoker Devices*	
a. Stationary grates, bar and pingrates	a. Combustion air needed to keep grates cool, causes fly ash entrainment. Problems with heat recovery surfaces and emissions. Usually not suited for continuous operation (except for low ash products, like wood).
b. Movable grates, chain grate, rocker grate, roller grate	b. Need alloy-type grates to avoid burn-out problems due to dripping plastic, tar, rubber, etc.; also to avoid corrosion from combustion. Abrasion and corrosion of the many parts necessitates maintenance shutdowns and replacements.
c. Rotary drums, refractory cylindrical & conical	c. Successfully used as conveying chamber. Typically preceded with movable grates on municipal wastes. Used for burning sludges. High cost typically requires larger installations to justify. High particle entrainment due to high gas velocities create problems for heat recovery surfaces and emission control devices.
d. Rotary drums, water tube	d. New approach. Used primarily in municipal wastes. No history of abrasion, hydrocarbon corrosion in metal tubes of drum or other types of corrosion.
e. Stationary hearth	e. Typically not used for 24-hour operation. Use for pathological wastes and batch feeds not exceeding 10 hours/day reduces heat recovery capability. Refractory slagging is main difficulty.
f. Stationary hearth with transfer rams	f. Used to achieve 24 hour/day operation. Typically used with controlled air furnaces. Main problem: fixed carbon loss of Btu's in waste or if additional combustion air is used suffers slagging of floors and/or ram maintenance. Moving rams break furnace seals and nullify controlled combustion. Possible fire hazards of plastic and oily

(continued)

TABLE 3.11
(continued)

System Components	Potential Problems; Operating Experiences
	materials dragged out by retracting rams. Fires have caused extensive damage to systems equipment outside of furnace.
g. Fluidized Bed	g. To date has not been successful in burning industrial wastes except for wood, coal, and sludges (not contaminated with heavy metals). The system can burn processed industrial wastes but emissions from bed create excessive demands on pollution-control equipment to remove extremely fine particulates and/or contaminate heat transfer surfaces and require excessive downtime for maintenance.
h. Pulse Hearth	h. Burns broad spectrum of Btu/lb. wastes either in bulk or processed. Has advantages of stationary hearth which prevents granular materials from falling through, or plastics and tars from dripping through grates. The pulse action crevices the firebed and moves it along like movable grates, without costly moving parts. Initial units had design problems with suspension system, furnace seals, and pulse actuating components. These problems have been de-bugged and redesigns show low maintenance costs and improved on-time for the system.
4. *Heat Capturing Surfaces*	
a. *Main Chambers*	
(1) Refractory furnace walls	(1) No capture of combustion heat. Heat loss from chamber is a function of insulation and refractory thicknesses. Smaller industrial units tend to be compromised on price, causing quicker breakdown of refractories. Units run 300 to 400 degrees F. hotter in chamber than water wall units. Extra heat is a factor in shorter service life to refractories. Up and down temperatures with non-continuous operating units add to maintenance problems. Excess air furnaces run cooler but add to

(continued)

TABLE 3.11
(continued)

System Components	Potential Problems; Operating Experiences
	emission problems and heat recovery surface problems.
(2) Water Walls integral design	(2) Most European and American water-wall boilers have the main radiant chamber and convection chamber closely integrated. Water walls must be protected from carbon corrosion of metals and reactions with chlorine and sulfur compounds. Refractory has been used successfully to cover areas of reducing atmosphere corrosion. Proper steam pressures with continuous use reduce acid dew-point corrosion. Absence of controlled-air tunnels due to integral radiant and convection design prevents this kind of boiler from burning hot industrial waste Btu's without smoking. Most successfully used on lower Btu municipal wastes or wood wastes.
(3) Water Walls with controlled air design	(3) A new (patents applied for) use of membrane water wall in a controlled air concept is being successfully utilized to obtain advantages of burning hot industrial waste Btu's and generating steam efficiently with the use of the radiant membrane water wall. Successful history of burning a variety of wastes without problems to the membrane water wall over a three-year period. Such items as asphalt shingles, PVC plastics, poly plastics, plastic foams, woods up to 50% moisture, sludges, and municipal wastes have been burned. A special film coating is painted on to protect the metal surfaces. Longest use of three years shows no need for recoating.
(4) Rotary Water Wall	(4) Used by one vendor, also as a proprietary system. Only use to date has been on municipal wastes. Has history of approximately six months on stateside wastes.
b. *Convection Section*	
(1) Water tube boiler, Integral Design	(1) System of bare tubes employed with proper spacing to avoid fly ash build-

(continued)

TABLE 3.11
(continued)

System Components	Potential Problems; Operating Experiences
	ups in tube sections. Most systems which operate up to 700 degrees F. metal temperatures do not suffer corrosion. Above that temperature, system suffers gas phase corrosion from chlorine molecules. Design controls boiler slagging of tubes with lower temperatures reaching the gas passages of the tubes. General industrial wastes of high Btu's not generally applied. Most experience is on municipal wastes.
(2) Water Tube Boiler with controlled air design using membrane water walls	(2) Three year history shows no slagging of tubes when surfaces are bare tubes. One finned-tube installation burning PVC materials had buildups which required downtime weekly for washing off. Conventional air or steam soot-blowers keep bare tube surfaces clean.
(3) Fire Tube Boiler with Controlled Design Using Membrane Water Wall	(3) Fire tubes show equally good history. Cooler burning in main chamber prevents slagging of tubes. Corrosion due to acid dew point can be controlled with proper steam pressure selections. Design limited to 300 psi because of drum costs.
(4) Water Tube Recovery with Incinerators	(4) Not enough history with industrial wastes (which would create slags on tubes). Design must be very selective. Bare tubes are essential. Cleaning of potential slagged surfaces must be addressed because temperatures can run away easier in a refractory-lined main combustion chamber when one is adding undetermined amounts of Btu's of non-homogeneous wastes. Finned-tubes have been used for several years by one vendor whose design offers removable banks of fined tube sections of easier cleaning. However, stopping to remove the tube backs to cool down, remove and plug and re-heat, cuts into available time. Selection of all bare tubes (water tube or fire tube) may have larger capital cost, but could be offset with greater energy savings.

(continued)

TABLE 3.11
(continued)

System Components	Potential Problems; Operating Experiences
(5) Fire Tube with Incinerator Main Chamber	(5) Most commonly used design by the bulk of the incinerator vendors. Easier to clean manually than water tubes. Sootblowers available for "on the run" cleaning. If slag develops in tubes, rodding shortens tube life. Limited to 300 psi because of drum costs.
(6) Economizers	(6) The selection of the economizer should be dependent on the type of wastes burned. Finned-tube types are less costly than bare tubes. However, bare tube costs can be offset by avoidance of plugging problems. Dew-point of gases should be carefully observed so that acids do not corrode the economizer. Material selection of costlier alloys may be offset with greater heat recovery. Systems that settle for a "partial loaf" and exit 350 Degrees F. and above, are recommended with typical carbon steel construction.
5. *Ash Removers*	
a. Drag Bar Type	a. Conventional ash removing has drag chains "living under the water," dragging the ash out of typically water-sealed furnace ash pits. These are usually high-maintenance items with initial expense. The drag-bar system is a carryover from furnace designs that used water under the grates so that live ash dropping through could be quenched and dragged or washed toward the ash pit area.
b. Scoop Type Ash Systems	b. There are some new proprietary designs of ash removal devices using the back hoe or scoop mechanism to pick the ash out of the ash pit. They lend themselves to low maintenance and easy access for lubrication since the only items in the ash pit are ashes and water.

REFERENCES

1. AP-42, "Compilation of Air Pollutant Emission Factors," Third Edition, U.S. EPA OAQPS (January 1984).
2. McIlvaine, Robert (ed.), "The Electrostatic Precipitator Manual," The McIlvaine Co., Chapter 9 (1976).
3. Hesketh, Howard E., "Air Pollution Control," Ann Arbor Science/Butterworths Publishers, second printing (1981).
4. Kunzweiler, V. L. and Vineyard, C. S., "Stack Testing of High Temperature Incinerators," APCA Technical Conference on the Effect of Disposal of Hazardous Waste on Air Quality, Kansas City, MO (February, 1981).
5. Olexsey, R. A. and Mournigham, R. E., "Emissions Testing of Industrial Processes Burning Hazardous Waste Materials," 11th ASME National Waste Processing Conference (1984).
6. Polcyn, A. J. and Hesketh, H. E., "Sampling and Analytical Methods for Assessing Toxic and Hazardous Organic Emissions From Stationary Sources," *Journal of the Air Pollution Control Association*, Vol. 35, No. 1, pp. 54–60 (January 1985).
7. Air Pollution Control Association, Proceedings—APCA Specialty Conference on, "Measurement and Monitoring of Non-Criteria (Toxic) Contaminants in Air" (March 22-24, 1984).
8. Allie, S. and Ranchoux, R., "Stack Sampling of Organic Compounds. Application to the Measurements of Pollution Control Devices Efficiency." Paper Presented at 73rd Annual Meeting of APCA in Montreal, Canada (June 1980).
9. Guzewich, D. C., Bond, C. A., Valis, R. J., Vinopal, J. H. and Deter, D. P., "Air Sampling Around a Hazardous Liquid Surface Impoundment." Paper presented at 76th Annual Meeting of the APCA in Atlanta, Georgia (June 1983).
10. Haile, Clarence L. and Baladi, Emile, "Methods for Determining the Total Polychlorinated Biphyl Emissions from Incineration and Capacitor—and Transformer Filling Plants." EPA Contract No. 68-02-1780, Task 2.
11. Hijazi, N. H., Chai, R. and Nacson, S., "A Methodology for Monitoring Air Pollutants on Industrial Landfill Sites." Paper presented at 75th Annual meeting of the APCA in New Orleans, Louisiana (June 1982).
12. Howe, Gary B., Gangwal, S. K. and Jayanty, R. K. M., "Validation of EPA Reference Method 25—Determination of Total Gaseous Nonmethane Organic Emissions as Carbon." Paper presented at 75th Annual Meeting of APCA in New Orleans, Louisiana (June 1982).
13. Johnson, L. "Protocol for the Collection and Analysis of Volatile POHC's using VOST." Draft document prepared by Envirodyne Engineers, Inc. for U.S. EPA/IERL-RTP (August 1983).
14. Jungclaus, G. and Garman, P., "Draft Final Report, Evaluation of a Volatile Organic Sampling Train," Midwest Research Institute, EPA Contract No. 68-01-5915 (1982).

15 Keith, Lawrence H., "Identification and Analysis of Organic Pollutants in Air" (1984).
16 Lane, Douglas A., "Mobile Mass Spectrometry—A New Technique for Rapid Environmental Analysis." Paper published in Environmental Science and Technology. Volume 16, No. 1 (1982).
17 Levins, P. L., Rechsteiner, C. E. and Stauffer, J. L., "Measurement of PCB Emissions from Combustion Sources" (1979).
18 Rechsteiner, C., Harris, J. C., Thrun, K. E., Sorlin, D. J. and Grady, V., "Sampling and Analysis Methods for Hazardous Waste Incineration," Arthur D. Little, Inc., EPA Contract No. 68-02-311, EPA/IERL, RTP, NC (1981).
19 Reinke, James M. and Devlin, Robert W., "Intercomparison of the Total Carbon Analysis, Flame Ionization Detection, and Gas Chromatographic Methods for Measuring Solvent Vapor Emissions." Paper presented at 73rd Annual Meeting of APCA in Montreal, Canada (June 1980).
20 Sakuma, T., Davidson, W. R., Lane, D. A., Thompson, B. A., Fulford, J. E. and Quan, E. S. K., Sciex, Inc. "The Rapid Analysis of Gaseous PAH and Other Combustion Related Compounds in Hot Gas Streams by APCI/MS and APCI/ MS/MS" (1982).
21 Scheil, George W. and Bergman, Frederick J., "Evaluation of a Volatile Organic Carbon Analyzer for Measuring Stack Emissions." Paper presented at 73rd Annual Meeting of APCA in Montreal, Canada (June 1980).
22 Stanley, J. S., Haile, C. L., Small, A. M. and Olson, E. P., "Sampling and Analysis Protocol for Assessing Organic Emissions from Stationary Combustion Sources in Exposure Evaluation Division Combustion Studies," Midwest Research Institute, EPA contract No. 68-01-5915 (1982).
23 U.S. EPA Office of Solid Waste and Mitre Corporation, "Guidance Manual for Hazardous Waste Incinerator Permits," Publication SW-966, EPA Contract No. 86-01-0092 (1983).
24 U.S. EPA, "Method 25—Determination of Total Gaseous Nonmethane Organic Emissions as Carbon," 40 CFR 60, Appendix A (revised as of July 1, 1983).
25 U.S. EPA, "Test Methods for Evaluating Solid Waste—Physical/Chemical Methods," SW-846, 2nd Edition (1982).
26 U.S. EPA, "Emissions of Reactive Volatile Organic Compounds From Utility Boilers," Document prepared by TRW, Inc. for U.S. EPA/IERL-RTP (May 1980).
27 U.S. EPA, OAQPS, "Measurement of Volatile Organic Compounds" (1979).
28 U.S. EPA, "Manual of Analytical Methods for the Analysis of Pesticide Residues in Human and Environmental Samples" (1977).
29 IERL-RTP Procedures Manual: Level 1 Environmental Assessment (2nd Edition), EPA 600/7-78-201 (October 1978).
30 U.S. EPA, "EPA Test Methods for Organic Chemical Analysis of Municipal and Industrial Wastewater," EPA-600/4-82-057 (July 1982).

31 Hofmann, R. E., "Refuse-to Energy in Less Than Two Years," *Public Works*, pp. 72–74 (March 1983).
32 Novak, R. G., Troxler, W. L. and Dehnke, T. H., "Recovering Energy From Hazardous Waste Incineration," CE, Vol 91, No. 6 (March 19, 1984).
33 Hesketh, H. E. and Cross, F. L., "The Pitfalls of Using Incineration for the Recycling of Drums Containing Chemicals or Hazardous Materials," AIChE Summer National Meeting, paper no. 65e, Philadelphia (August 1984).
34 Basic, John N., "Obtaining Consistant and Predictable Heat Recovery From Industrial Incineration Depends on Design Selections," 76th Annual APCA Meeting, paper no. 83-59.3, Atlanta (June 1983).
35 Carl, B. R., "Recovery Systems Reduce Hospital Energy Costs," *World Wastes*, pp. 14–15 (May 1983).

APPENDIX

FANS AND DAMPERS

Fans and dampers are used to control the air and gas flow within the cremator. Since they are an integral part of incinerator design, brief review of its various types is presented here.

Types of Fans

(a) *A straight radial blade fan* has four or more blades as shown in Figure A.1. It is low in both efficiency and cost, and is generally selected for slower speeds, although it can produce a very large tip speed.

(b) *The forward-curved blade fan* may have as many as 40–50 blades. It is generally similar to the straight radial blade fan in efficiency and application. It is somewhat more costly than a straight radial blade fan, but can be operated somewhat out of balance without damage. See Figure A.2.

(c) *The radial-tipped blade fan* is more efficient than either a straight radial or forward-curved blade fan and is shown in Figure A.3. It is often selected as an induced draft fan. It may have as many as 32 or 34 blades. This type requires a larger housing than a straight radial and costs more than either a straight blade or a forward-curved radial. However, it is quieter and easier to keep clean since it has a tendency to throw off fly ash. Since the buildup tends to be relatively even, the fan has less tendency to imbalance.

(d) *The backward-inclined flat blade fan* can only be used when the incoming air is clean (Figure A.4). Buildup of material at Point A, for instance, will cause serious imbalance which would probably destroy the fan. The fan has a fairly high efficiency.

(e) *The backward-inclined airfoil blade fan* has a high efficiency and

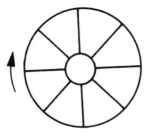

FIGURE A.1 Radial blade fan.

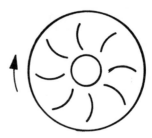

FIGURE A.2 Foreward blade fan.

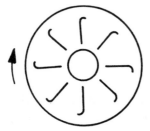

FIGURE A.3 Radial-tipped blade fan.

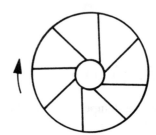

FIGURE A.4 Backward-inclined flat blade fan.

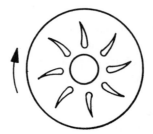

FIGURE A.5 Backward-inclined airfoil blade fan.

high cost. Eight to 10 or more airfoil blades may be used as in Figure A.5. Although this type of fan is seldom used for induced draft, it is very often used for forced draft. Again, the air must be clean.

Flow Volume Control

The control of the volume of gas from a fan can be controlled either by fan speed or by dampers. Fan speed control is more effective, but also more expensive. Thus, in most installations, volume is controlled by dampers.

(a) *A parallel blade damper* tends to direct the air to one side of the duct as it closes. It can completely shut off the flow. This is shown in Figure A.6.

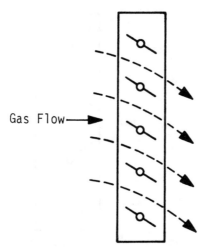

FIGURE A.6   Parallel blade damper.

(b) *An opposed blade damper* produces a straight flow, providing equal distribution downstream in the duct. This type can also shut off the flow completely as in Figure A.7.

(c) *An inlet valve* can control the flow to approximately 25% of the wide-open volume, with leakage through center only, Figure A.8.

Forced Draft Fan Systems

A forced draft system is one in which the fan is at the inlet end of the system, and blows or forces the air into the system. Forced draft fans have either the backward-inclined, flat-blade wheel, or the backward-inclined, air foil blade wheel. In general, they have ten to twelve blades, and a non-overloading characteristic in their horsepower curves.

Induced Draft Fan Systems

An induced draft system is one in which the fan is downstream of the inlet, and draws air into the system. When the gas to be handled is unlikely to be clean, the best choice is the radial-tip fan.

Quick-Draft Fan

Another method presently being used for incinerators is the quick draft type of fan. See Figure A.9.

Installation and Maintenance

Fans must be straight and level and all anchor bolts tightened *evenly*. Most complaints of fan vibration can be traced to improper installation.

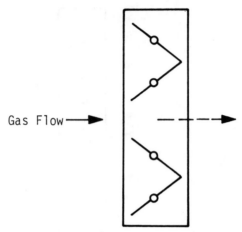

FIGURE A.7 Opposed blade damper.

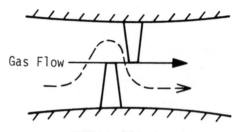

FIGURE A.8 Inlet valve.

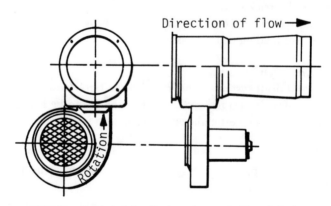

FIGURE A.9 Quick-draft fan (horizontal or vertical installation).

Both forced and induced draft fans must be cleaned as completely as possible, including all blades, to prevent the development of an unbalanced condition.

In general, the only other maintenance a fan requires is at the bearings, or grease fittings on dampers and vanes.

CONVERSION FACTORS

Approximate values for incinerator flue gas dusts (assuming: standard heat values; given emission factors; that the flue gas = dry air; and corrected to 50% excess air $\cong$ 12% CO_2).

Multiply	By	To Obtain
grains/scf	1.90	lb/1000 lb flue gas
grains/scf	2.20	lb/10^6 Btu heat input
lb flue gas/hr	0.221	scfm
lb/100 lb Type 0 Waste	0.532	grains/scf
lb/100 lb Type 1 Waste	0.685	grains/scf
lb/100 lb Type 2 Waste	0.970	grains/scf
lb/100 lb Type 3 Waste	1.475	grains/scf

LOWER EXPLOSIVE LIMITS IN AIR

Material	LEL, % by Vol.
Carbon Monoxide	12.5
Methyl Alcohol	7.3
Methane	5.2
Acetaldehyde	4.0
Hydrogen	4.0
Ethylene	3.1
Methyl Acetate	3.1
Acetone	3.0
Ethane	3.0
Acrolein	2.8
Propylene	2.4
Furfural	2.1
Isopropyl Alcohol	2.0
Butane	1.9
Methyl Ethyl Ketone	1.8
Cellosolve Acetate	1.7
Benzene	1.4
Toluene	1.4
Cyclohexanone	1.1
Styrene	1.1
Xylene	1.0

INDEX

Air Control	41		*Material*	92
Air Control, Manual	42		Equivalence Ratio	15
Air Control, Temperature Modulated	42		Excess Air	14,35
Air-Fuel Ratio	14		Fans	107
Analytical Methods	87		*Flow Control*	108
Behavior	41		*Installation*	109
Burners	53		*Maintenance*	109
controls	56		*Types*	108
gas	54		Feed	35
gas adjustments	63		Flue Gas, Composition	18
gas-oil	55		Flue Gas, Quantity	20
oil	54		Hazardous Wastes	74
rate checks	57		Heating Values	6
safety	56		Ignition Systems	55
Carbon, Fixed	8		Incinerator, Temperature, Color	45
Charging	35		Loading (Charging)	35
Clinkers	47		Maintenance	61
Combustible Wastes	5		Moisture	8
Combustion	9		Non-Combustibles	8
Combustion, Factors of	15		Operation	65
Combustion, Simplified	16		*Emergency*	66
Controlled Air Incinerators	3		*Shut Down*	66
Conversion Factors	111		Operational Problems	46
Costs			Organic Emissions—Sampling	
Air Pollution Control	96		Procedures	75
Incineration Systems	95		Particle Emissions	
Dampers	107		*Control*	71
Design Considerations	58		*Testing*	73
Devolatilization	16		*Size Distribution*	71
Emissions	69,75		Quality Control	87
Factors	71		Reclamation Furnaces	92
Emissions—Analytical Methods	83		Refractory	
Emission Sampling Methods	76		*Chemical Thermal and Corrosion*	
Energy Recovery	88		*Concerns*	49
Heat	89		*Inspection*	63

113

Installation	53	Temperature Control	35
Selection	50	Trouble Shooting	63
Types	48	*Blower*	64
Sampling	75	*Burner*	64
Sampling and Analytical Quality Control	87	*Oil Burner*	64
		Volatile Matter	7
Spare Parts	57	Waste	59
Temperature, Color	45	Wastes, Combustible	5

BIOGRAPHIES

Frank Cross

Frank Cross has over 25 years of diversified environmental engineering experience in industry and in government. Emphasis in air pollution control has been on ambient and source testing, concept engineering and regional air quality management (modeling).

Mr. Cross has worked on environmental impact statements which have involved air pollution control, water pollution control, and noise abatement concepts. Hazardous waste management is a specialty. He has been a professional trainer for many years and was former Deputy Chief of the Training Program for the EPA.

Mr. Cross has Bachelor of Science Degrees in Chemical and Civil Engineering, and a Master of Engineering Degree in Air Pollution Control. He is a registered Professional Engineer in ten states and is a Diplomate of Air Pollution Control in the American Academy of Environmental Engineers.

Mr. Cross was a Principal Consultant with Roy F. Weston, Inc., Consulting Engineers and Scientists, and Operations Field Services Administrator for the State of Florida Department of Pollution Control. Presently, he is President of Cross/Tessitore & Associates, P.A.

Howard E. Hesketh

Howard E. Hesketh is a Ph.D. chemical engineer and a professor at Southern Illinois University-Carbondale, where he instructs courses in air pollution control, engineering economics and thermodynamics. He has worked extensively with industry, having served as a consultant to a number of industrial facilities as well as to U.S. Federal and local governmental agencies, foreign government agencies and other organizations. He is a registered Professional Engineer and a certified air pollution control engineer by the American Academy of Environmental Engineers. Dr. Hesketh is author of three engineering texts, has co-authored several books, is Associate Editor of the journal, Atmospheric Environment and is Associate Technical Editor of Journal of Engineering for Power. He currently is a member of the Board of Directors of the Air Pollution Control Association and is a member of the Board of Trustees of the American Academy of Environmental Engineers.